CONTEMPORARY

NUMBER
POWER

A REAL WORLD APPROACH TO MATH

Analyzing Data

ELLEN CARLEY FRECHETTE

D0902772

Mc Graw Hill **Education**

Bothell, WA • Chicago, IL • Columbus, OH • New York, NY

www.mheonline.com

 **Education**

Send all inquiries to:
Contemporary/McGraw-Hill
130 E. Randolph, Suite 400
Chicago, IL 60601

ISBN: 978-0-07-659233-3
MHID: 0-07-659233-2

Printed in the United States of America.

2 3 4 5 6 7 8 9 RHR 15 14 13 12 11

TABLE OF CONTENTS

Analyzing Data

Using Probability and Statistics

Evaluating Data

Posttest A   134

USING NUMBER POWER

TO THE STUDENT

Do the following statistics sound familiar?

- There's a 40% chance of rain.
- The unemployment rate is 4%.
- 3 out of 4 people prefer brand X.
- $3 million will be added to the district education budget.

Welcome to *Number Power Analyzing Data*. With this book, you will learn the language of data and statistics so that you can analyze and understand the numbers around you. This understanding will help you feel more comfortable and knowledgeable when faced with statistics.

Begin to pay more attention to the numbers around you. Find some statistics in a newspaper article or editorial. Think about how you might interpret them *differently*. Remember that although numbers may be based on facts, the statements made about them are not always accurate. *Analyzing Data* will help you understand how to interpret data and how to judge other people's interpretations.

The first part of the book, Building Number Power, contains five chapters giving you step-by-step instruction and practice with various kinds of data and statistics. The second part of the book, Using Number Power, gives you a chance to apply and extend the data analysis skills you have just learned.

As you work through *Analyzing Data*, check your answers with the answer key at the back of the book. A chart inside the back cover will help you keep track of your scores.

Analyzing Data Pretest

This test will tell you which sections of this book you need to concentrate on. Do every problem that you can. Correct answers are listed by page number at the back of the book. After you check your answers, the chart at the end of the test will guide you to the pages of the book where you need work.

For problems 1–4, refer to the following data.

Projected Population of Selected Large Cities in 2020	
City	2020 Population (to the nearest million)
Tokyo, Japan	37,000,000
Mumbai, India	26,000,000
Dhaka, Bangladesh	22,000,000
New York, USA	20,000,000
Shanghai, China	13,000,000

Source: City Mayors Statistics, 2010

1. Which of the cities listed will have the largest population?

   a. Shanghai

   b. Tokyo

   c. New York

2. Complete the following sentence.
   The table above shows the projected 2020 _____ to the nearest million for five cities.

3. Write a sentence that compares the expected 2020 population of Dhaka to the expected 2020 population of New York.

4. How many more people are expected to live in Tokyo than in Mumbai in 2020?

   a. 37,000,000

   b. 26,000,000

   c. 11,000,000

For problems 5–8, refer to the following data.

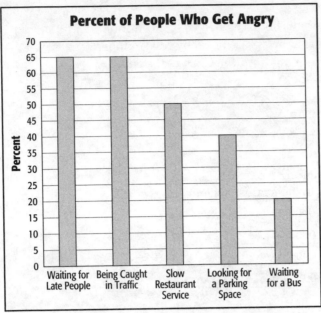

Source: Almanac of the American People

5. Approximately what percent of people get angry looking for a parking space?

   a. 40%

   b. 50%

   c. 55%

6. Which situation made more people angry: being caught in traffic or waiting for a bus?

7. Write a statement about the percent of people who get angry because of slow restaurant service.

8. Find the mean (average) percent of people who get angry in the five situations on the graph.

For problems 9–12, refer to the following data.

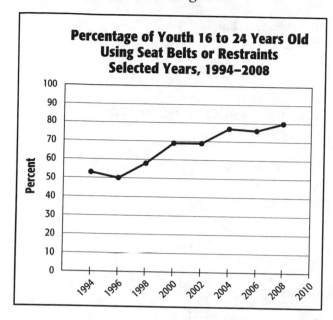

Percentage of Youth 16 to 24 Years Old Using Seat Belts or Restraints Selected Years, 1994–2008

9. What percent of youth 16–24 years old used seatbelts or retraints in 2004?

   a. 69%

   b. 74%

   c. 77%

10. Estimate and compare the percent of youth 16–24 using seatbelts or restraints in 1998 and 2008.

    In 1998, approximately _____ percent fewer youth 16–24 used seatbelts than in 2008.

    a. twenty

    b. thirty

    c. forty

11. If the trend continued, what percent of youths 16–24 used seatbelts or restraints in the year 2010?

    a. about 80%

    b. more than 80%

    c. less than 80%

12. Which of these statements is *not* based on the graph?

    a. More youths age 16–24 used seatbelts or restraints in 2008 than in 2006.

    b. In general, the percent of youths age 16–24 who wear seatbelts has risen over time.

    c. There were fewer automobile deaths among youths 16–24 in 2008 than in prior years.

For problems 13–15, refer to the following data.

According to the U.S. Department of Housing and Urban Development, 37% of homeless people in 2009 were families. Single men made up 43% of the homeless population, and 20% of the homeless were single women. These statistics provide more evidence of the changing face of America's homeless.

13. Complete the circle graph below. Be sure to include all labels and numbers for each section of the graph.

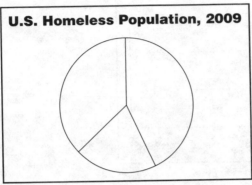

U.S. Homeless Population, 2009

Source: U.S. Department of Housing and Urban Development

14. If an average city counted 1,200 homeless people, how many of them would be single women?

    a. 1000

    b. 480

    c. 240

15. Write a statement that compares two sections on the circle graph.

For problems 16–17, refer to the following situation.

The cards at the right are lying facedown on a table.

| 1 | 2 | 3 | 6 | 10 |

16. Suppose you pick up one card. What is the probability that the card you chose is a multiple of 2?

    a. 1 out of 2

    b. 2 out of 3

    c. 3 out of 5

17. What is the probability of choosing a 5 from these cards?

    a. 0

    b. $\frac{1}{2}$

    c. 1

For problem 18, refer to the following information.

A year 2010 survey indicated that most Danforth County residents consider themselves to be overweight. A total of 80% of people surveyed feel that they should lose at least 10 pounds. These figures show an increase from previous years. In 1980, for example, 59% of people surveyed felt that they were 10 pounds overweight. In 1990, the figure rose to 72%. And in 2000, the figure was 78%.

18. Use the data above to finish constructing the line graph below. Be sure to include a title, labels on both axes, and the data line connecting the plotted points.

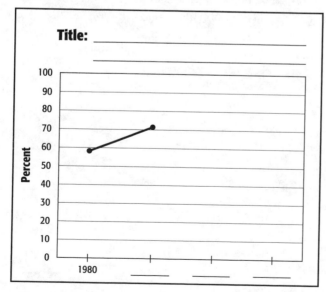

For problem 19, refer to the following scatter diagram.

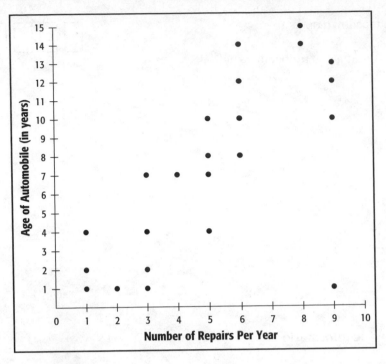

19. Write a statement comparing the age of automobiles and the number of repairs per year.

_____

ANALYZING DATA PRETEST CHART

If you miss more than one problem in any section of this test, you should complete the lessons on the practice pages indicated on this chart. If you miss no problems in a section of this test, you may not need further study in that chapter. However, to master data analysis, we recommend that you work through the entire book. As you do, focus on the skills covered in each chapter.

Problem Numbers	Skill Area	Practice Pages
1, 2, 4	tables and charts	26–29
3, 7, 15	making statements about data	11–12
5, 6	bar graphs	30–35
8	mean, median, and mode	52–56
9, 18	line graphs	36–40
10	estimation	69–73
11	seeing trends	82–85
12	using only the information given	66–68
13, 14	circle graphs	44–46
16, 17	probability	86–92
19	understanding correlation	119–123

BUILDING
NUMBER
POWER

UNDERSTANDING DATA

Defining Data

Let's start our discussion of data with questions about you.

- What is your age?
- How many people are in your immediate family?
- How many people are living in your home?
- How many languages do you speak?
- How many times per week do you read the newspaper?
- How many different sports do you watch regularly on TV?

The numbers you gave in answer to the questions above represent **data.** *Data* refers to a collection of numbers that gives information about a subject—in this case, you!

Here are some examples of data that are common in everyday life. Put a check next to any you are familiar with.

The nutrition label on a cereal box gives data on calories, fat content, vitamins, and sodium.

NUTRITION INFORMATION		
CALORIES		160
PROTEIN, g		4
CARBOHYDRATE, g		33
FAT, TOTAL, g		3
UNSATURATED, g	3	
SATURATED, g	0	
CHOLESTEROL, mg		0
SODIUM, mg		40
POTASSIUM, mg		125

A graph in a newspaper displays population data.

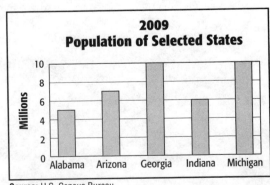

A document for factory employees contains data about the sizes of certain manufactured items.

Identification Number	Size
1A714	12 mm
1A820	14 mm
2B640	20 mm
2B724	26 mm
3C438	30 mm

An article in a magazine provides data concerning the use of recycled products.

People use data all the time to analyze situations and make decisions. Chances are that you are familiar with at least one of these examples of data.

A recent survey of Rand County residents conducted by Environment Aware, Inc., revealed that, of those surveyed,

- 47% buy and use recycled products.

- 90% would buy more recycled products if the prices were lower.

- 76% have trouble finding recycled products.

Gathering Data

Where do we get data? How is it collected? Depending on the subject, data can be gathered in many different ways. For example,

- A scientist might collect data by measuring the quantities of chemicals in a series of test tubes.

- A teacher might collect data about test scores by keeping a record of all test and quiz grades given in a class.

- An environmental worker might gather data by observing and recording what kinds of animals and birds live in a certain region.

- A government worker might collect data by analyzing questionnaires filled out by citizens.

In fact, you may not realize it, but various types of data are being collected around you every day. If you have ever filled out a census form or responded to a telephone survey, you have supplied data that some person or organization is collecting.

Check off any of the forms of data collection that you have participated in.

questionnaire
consumer product survey
credit or job application form
U.S. Census form
opinion poll
voting
other (What was it? _____)

To get a better feel for data, collect your own data in this activity.

Choose a street corner near where you live and record information about the first 10 people you see, using the chart below. Or, observe the first 10 characters you see on a TV program. To record the data when you see a person, put a slash, or **tally mark (|)**, in the columns that apply. Tally marks help you "see the numbers."

(**Hint:** When you have recorded four tally marks in a row, your *fifth* mark should be a slash through the four marks, like this:

This tallying method makes it easier to count up your marks later.)

For example, if the first person you saw was a frowning man wearing glasses and carrying a shopping bag, your chart would look like this:

Male	Female	Wearing Glasses	Not Wearing Glasses	Smiling	Not Smiling	Carrying Something	Not Carrying Anything
|		|			|	|	

Making Statements About Data

Fall 2008 Enrollment in Full-Time Day Schools		
State	Pupils per Teacher	Teachers' Average Pay
Alabama	14.8	$48,233
Alaska	17.2	$58,105
California	20.8	$66,064
New Jersey	12.4	$62,150

Source: National Center for Education Statistics

Write two short statements using the data in the chart. There is no single correct answer, so feel free to be creative. The first one is started for you.

1. California has _____

_____

2. _____

_____

What can be said about the information in the chart? Here are some statements you might have written about the data in the chart.

- California has more pupils per teacher than Alabama.
- Of the four states listed, California has the highest teachers' average pay.
- Alabama teachers have the lowest average pay of the four states listed, but not the fewest pupils per teacher.
- Teachers' salaries vary from state to state.

Some of these statements refer to specific pieces of information in the data.

EXAMPLE   California has more pupils per teacher than Alabama.

Other statements make a more general comment about the data.

EXAMPLE   Teachers' salaries vary from state to state.

The one thing that these statements have in common is that they all say something *true* about the data.

Throughout this book, you will be looking at different types of data and making statements about them. Don't worry if you had trouble coming up with a statement about the data in the table. You will get plenty of practice. By the end of this book, you will be able to *interpret and write statements* about almost any data you come across in school, at work, or in a newspaper or magazine.

IMPORTANT!

Whenever you make a statement about data, be sure to write a **complete sentence**. The phrase *seventeen pupils per teacher* does not have any meaning. However, *Alaska averaged about seventeen pupils per teacher in 2008* does have meaning.

Practice writing statements about the data you collected on page 10. The questions below should help you get started. Remember that you can be as straightforward or creative as you like when writing your statements. Be sure that each statement is a complete sentence and is true according to the data you collected.

Below are some questions to help you get started.

- Out of 10 people, how many were female?
- Did you see more people with glasses or without?
- How many people were carrying something?
- Who was smiling more often—men or women?
- Did more men than women wear glasses?

Statement 1: _____

_____

Statement 2: _____

_____

Statement 3: _____

_____

Statement 4: _____

_____

Statement 5: _____

_____

What Are Statistics?

Of the 4,280 people who responded to our magazine's survey, 71% said that the government should provide subsidized child care for all workers. Twenty percent stated that child care is not the government's concern, and 9% had no opinion on the issue.

What numbers appear in this article?

Have you read articles similar to this?

How do these numbers differ from the data you saw earlier?

Many statements you read in newspapers and magazines or hear on the radio or television contain **statistics** such as those in the article above. The term statistics refers to the collection, organization, and interpretation of data.

For example, on page 10, you collected data about a group of people. If you organized and interpreted this information, you might use statistics to help present the facts.

Imagine what the article above would sound like if it simply listed the information obtained from the survey.

QUESTION: Do you think that the government should provide subsidized child care for all workers?

Miriam Brown from Spokane, WA, responded yes.

Carlos Young from Cairo, IL, responded yes.

David Antony from Los Angeles, CA, responded no.

Walter Fireton from Baton Rouge, LA, responded yes.

Vera Roundtree from Orono, ME, responded no.

Mary Caviness from Kansas City, MO, had no opinion.

Maureen Burns from Windham, NH, responded yes.

⋮

(And the list continues on.)

As you can see, an article that listed each person's opinion would be extremely long. And it would be practically impossible to understand the results of the survey. Instead, the writer **summarizes** the data by using percents.

71% said "yes."

20% said "no."

9% said "no opinion."

Using tally marks, record the responses given on page 13 to help you "see the numbers." Miriam Brown's response has already been recorded.

Yes	No	No Opinion

How Are Statistics Presented?

Once a group of data is collected, the numbers are counted and put into a form that is easy for people to understand.

In the example on page 13, the magazine probably issued a questionnaire or did a phone survey to determine what people thought about government-subsidized child care.

- What different ways can you think of to organize the information?
- How did the magazine present its data?

Math Recap
Data can be represented by **percents.** Remember that *percent* means "out of one hundred."

The statement that 71% think government should provide child care means that 71 out of every 100 people surveyed think government should provide child care.

The magazine received 4,280 responses to its survey. It chose to represent its data as *percents*. What other methods could it have used?

1. The magazine could have *listed* each individual's response separately, as you saw on page 13. This is a long and awkward method.

2. The magazine could have given *numbers* instead of percents.

   "Out of 4,280 responses, 856 people did not like the idea of government-subsidized child care."

3. The magazine also could have used *fractions*.

   "Approximately $\frac{7}{10}$ of respondents favored government-subsidized child care."

4. *Ratios* are another way to present data. You may be familiar with this technique in advertisements.

   "7 out of 10 (7:10) respondents favored government-subsidized child care."

   Ratios are most often written in fraction form: $\frac{7}{10}$.

Read the following lines taken from newspapers, magazines, and work reports. Express the numbers given as a percent, a fraction ratio, and a ratio.

"3 out of 10 students do not own a calculator."

30 %	$\frac{3}{10}$	3:10 or 30:100
percent	fraction ratio	ratio

1. "Employees pass the physical exam 95 out of 100 times."

    percent   fraction ratio   ratio

2. "8 out of every 100 people in Newton are unemployed."

    percent   fraction ratio   ratio

3. "7 out of 10 customers prefer orange to grape juice."

    percent   fraction ratio   ratio

4. "Out of 200 questionnaires returned, 60 were completed by females."

    percent   fraction ratio   ratio

5. "The risk of mechanical error is 1 out of 100."

    percent   fraction ratio   ratio

6. "The ratio of computer users to non-users was 4 to 5."

    percent   fraction ratio   ratio

Comparing Numbers

- Your town government votes to give $80,000 to the parks and recreation department.

- You get an offer of $80,000 for your house.

- The legislature voted in favor of an $80,000 government-subsidized school lunch program.

Is 80,000 a large number or a small number?
How do you decide?

> One of the most important things to know about a number is this:
> A number is neither large nor small until you compare it to another number.

Let's look at the first example above. What other information would help you decide whether or not $80,000 is a lot of money for parks and recreation?

You might compare $80,000 to some of these numbers.

- **The amount of money allocated the previous year.** For example, if the town had allocated $200,000 last year, perhaps $80,000 does not seem like a very large figure.

- **The total amount of money in the town budget.** If the town had a budget of $125,000, then $80,000 would seem like a lot of money, given the fact that the town must also support schools, fire and police departments, maintenance work, and so forth.

- **The amount allotted to parks and recreation in similar neighboring towns.** If you compared $80,000 to the $10,000, $8,000, and $22,000 that three similar towns were spending, you would certainly think $80,000 was a lot of money.

When you see or hear a number, be careful not to make a judgment about its size until you have a chance to compare it to other related figures.

Which of the following numbers would be useful in determining whether the $80,000 offered for your house is a large or small number? Put a check mark next to the numbers that you would use to compare with the $80,000.

a. the amount paid for a similar house in your neighborhood

b. your neighbor's salary

c. the number of houses on your street

d. the price you paid for the house last year

You were right if you chose **a** and **d.** How high or low your neighbor's salary is or how many houses there are in your neighborhood will not tell you anything about the price of your house.

Now finish the following sentences.

- $80,000 for a house would be a *large* amount of money if

_____

- $80,000 for a house would be a *small* amount of money if

_____

Here are some sample sentences.

- $80,000 for a house would be a *large* amount of money if **I paid only $40,000 for it last year.**
- $80,000 for a house would be a *small* amount of money if **my neighbor sold the same kind of house for $100,000 two months ago.**

Make two statements about the $80,000 school lunch program.

1. $80,000 for the school lunch program would be a lot of money if

_____

_____

2. $80,000 for the school lunch program would be a small amount of money if

_____

_____

Don't rely on other people's interpretations of the size of numbers. It's important to draw your *own* conclusions about the information. A good start is to *compare* the information being presented to other amounts. To see how the same number can be viewed as both large and small, read the following excerpts from two different newscasts.

Yesterday, a crowd of approximately 100 people marched in front of City Hall, protesting the pay raises the legislature had voted in on Tuesday. A handful of people held up signs supporting the increased salaries.

Compared to last year's enormous protests over high salaries for government officials, the group of 100 or so protesters in front of City Hall yesterday was hardly noticed.

The *fact* is that 100 people stood outside City Hall. How you *interpret* this fact depends on what you compare the numbers to.

Look at the figures given below. What other numbers would be helpful in deciding whether these numbers are small or large?

EXAMPLE $100 to repair a stereo
It would be helpful to know *how much another place*

would charge for repairs.

3. 400 people at a fund-raiser
It would be helpful to know _____

_____

4. 90 people laid off in a company
It would be helpful to know _____

_____

5. 3 guns confiscated in a police raid
It would be helpful to know _____

_____

6. $20 to purchase a painting
It would be helpful to know _____

_____

7. $2,000 to buy a car
It would be helpful to know _____

_____

How Big Is a Billion?

Ticket sales from the movie *Avatar* exceeded $500 million in 2010.

The United States' federal debt (the amount by which government spending exceeds government income) in 2010 was greater than $8.6 trillion.

Do you have a good understanding of the size of the numbers above?
When you read a very large number, what do you do to determine its size?

For many people, the bigger a number is, the less meaning it has. Let's review **place value** to help in our understanding of large numbers.

1	one
$1 \times 10 = 10$	ten
$10 \times 10 = 100$	one hundred
$100 \times 10 = 1,000$	one thousand
$1,000 \times 10 = 10,000$	ten thousand
$10,000 \times 10 = 100,000$	one hundred thousand
$100,000 \times 10 = 1,000,000$	one million
$1,000,000 \times 10 = 10,000,000$	ten million
$10,000,000 \times 10 = 100,000,000$	one hundred million
$100,000,000 \times 10 = 1,000,000,000$	one billion

Express the values in word or numeral forms as needed.

EXAMPLES   30,000,000 *thirty million*

two trillion **2,000,000,000,000**

1. 900,000 _____

2. 8,000,000 _____

3. 90,000,000 _____

4. one hundred million _____

5. two billion _____

6. seven hundred thousand _____

Math Recap
To multiply a whole number by 10, add a zero at the right.

The function of zeros can help you understand the relative size of numbers.

EXAMPLE 1 One million is how many times greater than a hundred thousand?

STEP 1 Write each amount in numeral form. Put the lesser number on top.

100,000

1,000,000

STEP 2 Cross out one zero at a time in each number until there are no zeros left in the lesser number. Be sure to cross out the same number of zeros in each number.

1̶0̶0̶,̶0̶0̶0̶

1̶,̶0̶0̶0̶,̶0̶0̶0̶

STEP 3 Count the number of zeros left in the larger number.

1,0̶0̶0̶,̶0̶0̶0̶

One zero is left.

STEP 4 Your knowledge of place value tells you that one zero represents 10.

ANSWER: One million is 10 times greater than one hundred thousand.

EXAMPLE 2 How many times greater is one billion than one million?

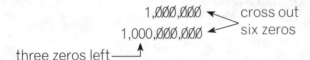

1,0̶0̶0̶,̶0̶0̶0̶ ← cross out six zeros

1,000,0̶0̶0̶,̶0̶0̶0̶

three zeros left

ANSWER: One billion equals one thousand times one million.

When you come across large numbers like these, you need to have a basic understanding of their *relative* size (size in relation to other numbers). There is a big difference in value between a million and a billion! To illustrate how big a difference, think about this:

- It takes $11\frac{1}{2}$ days for a million seconds to pass.
- It takes almost 32 years for a billion seconds to pass.
- If you had 5 million dollars, you could spend $10,000 a day for a year and still have more than a million dollars left.

Use the "times 10" chart on page 19 and the method shown on page 20 to fill in the following.

EXAMPLE Almost 4 million Roman Catholics live in East Asia. Almost 400 million Roman Catholics live in Latin America.

There are **_100_** times as many Roman Catholics in Latin America as in East Asia.

4,0̸0̸0̸,0̸0̸0̸   4 million

400,0̸0̸0̸,0̸0̸0̸   400 million
↗
two zeros mean
100 times

7. Approximately 160,000,000 Buddhists live in South Asia. About 160,000 Buddhists live in North America.

There are _____ times as many Buddhists in South Asia as in North America.

8. An estimated 750,000,000 people in East Asia consider themselves atheist or nonreligious. An estimated 75,000,000 Europeans include themselves in this category.

_____ times as many East Asians as Europeans consider themselves atheist or nonreligious.

9. About 100,000 Jews live in Oceania. About 1,000,000 Jews live in Latin America.

There are _____ times as many Jews living in Latin America as in Oceania.

10. Approximately 1,500,000,000 Christians exist worldwide, compared to 15,000,000 Jews.

There are approximately _____ times as many Christians as Jews worldwide.

11. About 90,000,000 Catholics live in Africa, out of a total of 900,000,000 Catholics worldwide.

There are _____ times as many Catholics in the world as in Africa.

Numbers vs. Rates

Which of the highways in the chart at the right would you prefer to drive on? Why?

2010 Accident Report	
Highway	Number of Accidents
A	110
B	75
D	10
G	90
HH	12
MM	170

Did you say that you would rather drive on *Highway D* because it had the fewest accidents and therefore was the safest to drive on?

You may have made the common mistake of confusing *numbers* with **rates**. Before you can decide which highway is the safest, you need to know some other numbers.

For example, would you change your mind if you knew that only 3,000 cars travel on Highway D per year, compared to 75,000 on Highway B?

The chart above really does not give you much useful information. What would be more helpful is to know how many accidents occurred per one hundred cars. In other words, you need to know the *rate* of accidents, not just the *number* of accidents.

Now look at the revised chart. Unlike the chart above, this chart gives enough information to determine the rate of accidents. Do you know how to determine the rates?

Highway	Number of Accidents	Total Number of Cars per Year
A	110	55,000
B	75	75,000
D	10	3,000
G	90	60,000
HH	12	8,000
MM	170	80,000

To figure out each rate, divide the number of accidents by the total number of cars.

Highway A: $110 \div 55{,}000 =$

A calculator entry would be

① ① ⓪ ÷ ⑤ ⑤ ⓪ ⓪ ⓪

The calculator displays

0.002

Change the decimal to a percent by moving the decimal point two places to the right.

$0.002 = 0.2\%$

Now find the accident rates for the remaining highways.

Which highway has the lowest accident rate?

Highway

A $\underline{0.002} = \underline{0.2}\ \%$

B   _____ = _____%

D   _____ = _____%

G   _____ = _____%

HH   _____ = _____%

MM   _____ = _____%

Highway B, with a rate of 0.1%, is the safest. Although this highway had 75 accidents, its accident *rate* is relatively low considering that so many cars travel on it.

Remember: A rate gives you more information than a number does, because a rate is actually *two* numbers—a number *compared to* another number.

A percent is the rate most commonly used—it represents a number out of one hundred. For example, which statement below gives you more information?

Statement A   There are 6,400 people living below the poverty level in our town.

Statement B   In our town, 17% of the people live below the poverty level.

Both statements are true, but statement B is the only statement that gives you information about the *size* of the poverty problem. Unless you know the total number of residents, statement A does not tell you much. The number 6,400 could be a relatively large number (if residents totaled 12,000) or a relatively small number (if residents totaled 220,000).

Look at the information below, then make

- **two statements about numbers**
- **two statements about rates**
- **a statement comparing numbers vs. rates**

A survey organization asked the employees of five different companies in the metro area this question: "Are you satisfied with your company's health benefits?" The results are shown below.

Company	Satisfied Employees	Total Number of Employees	Rate of Satisfaction
Walton Company	190	1,000	_____
Kramer, Inc.	13	87	_____
Psytech	440	3,800	_____
Ranger, Intl.	90	450	_____
HaML & Co.	26	105	_____

Listed below are some questions to help you get started.

1. What company had the *least number* of satisfied employees, and what is that number?

   Statement: _____

   _____

2. What companies had the *greatest number* of satisfied employees?

   Statement: _____

   _____

3. What company had the *lowest rate* of satisfaction?

   Statement: _____

   _____

4. What company had the *highest rate* of satisfaction?

   Statement: _____

   _____

5. Compare the *highest number* of satisfied employees with that company's employee satisfaction *rate*.

   Statement: _____

   _____

One of the most common rates that you will find in the news is the **unemployment rate.** Instead of reporting exactly how many people are out of work, the government reports a rate.

A 6.9% unemployment rate, for example, means that 6.9 (almost 7) people out of every 100 people in the labor force cannot find work.

The government also provides statistics for different segments of the labor force. For example, in 2010, the unemployment rate for the entire labor force was 9.5%. However, among young people ages 16–19, the rate was 26%—almost three times that rate!

Use the chart below to write three statements about the unemployment rates for various years and among various segments of the labor force.

U.S. Unemployment Rates			
Year	Total Workforce	White	Black
2005	5.1%	4.4%	9.5%
2006	4.6	4.0	8.4
2007	4.5	4.0	7.5
2008	5.4	4.9	8.9
2009	8.1	7.3	12.4

Source: Bureau of Labor Statistics

6. Compare the unemployment rate for the total work force in two different years.

_____

_____

7. Write a statement about unemployment among blacks in the United States using the figures above.

_____

_____

8. Compare the unemployment rates of whites to the rates of blacks during one year.

_____

_____

Tables and Charts

How Salty Snacks Stack Up (One-Ounce Serving)					
	Potato Chips	**Pretzels**	**Popcorn***	**Corn Chips**	**Peanuts**
Calories	152	111	76	153	170
Fat	10 g	1 g	1 g	9 g	14 g
From Fat	60% of calories	8% of calories	12% of calories	53% of calories	74% of calories
Sodium	160 mg	451 mg	1 mg	218 mg	138 mg

* Air-popped corn. The microwave type has about twice the calories. It contains 9 grams of fat (that's 55 percent of its calories) and 196 milligrams of sodium.

Source: Consumer Reports

Why are tables like the one above often used to display data?

What does this table have in common with other charts you've seen?

Charts and tables are useful because they organize data in an easy-to-read format. Effective tables and charts have at least four elements.

- A **title** describes the subject of the table or chart.

- **Headings** (**column** and **row** labels) show what information will be provided in the table.

- **Data** (for example, the numbers listed under and next to the headings above) gives the most specific information in the table.

- The **source** is usually found in smaller print below a graph, table, or chart, and it tells where the data actually came from.

To find specific information in a table, you will need to read across a row and down a column to where the row and column intersect (meet).

Look at the table on page 26. How much sodium is
in one ounce of corn chips?

STEP 1   Find *Sodium* in the list of row headings at the left in
the table.

STEP 2   Follow the row across until you come to the *Corn
Chips* column, labeled at the top of the table.

STEP 3   The number at the intersection of the *sodium* row
and the *Corn Chips* column is 218 mg.

ANSWER: There are **218 mg** of sodium in one ounce of corn chips.

Use the table on page 26 to fill in the blanks.

1. One ounce of air-popped popcorn contains _____ gram(s) of fat.

2. _____% of the calories in peanuts come from fat.

3. One ounce of microwave popcorn has about _____ calories.

Use the table to decide if each statement is true or false.

T F **4.**   Corn chips have more calories than potato chips.

T F **5.**   Popcorn has 4% more calories from fat than pretzels.

T F **6.**   A bowl containing one ounce of corn chips and one ounce of
pretzels contains more than 200 calories.

**Make two statements using the data in the table. The first one has been
started for you.**

7. A one-ounce serving of peanuts _____

_____

8. _____

_____

9. **Critical Thinking** *Consumer Reports* is a magazine published by a nonprofit,
independent organization. Do you think the information in its charts would be
different from charts you might see in advertisements? Why?

Use the information given in this newspaper article to fill in the table below.

STEP 1 Read the article below.

STEP 2 Write a title for the table and fill in the source line.

STEP 3 Label the columns (*$50,000 or More* has been filled in)
and rows (the *Bank Credit Card* row has been labeled).

STEP 4 Fill in the data (*94%* has been filled in).

According to *Money Express* magazine, 94% of Americans earning $50,000 or more had bank credit cards, and 54% had gasoline credit cards. Fourteen percent of this group had an Express Gold card. Of Americans earning $25,000–$34,999, 73% had a bank credit card, 33% had gasoline cards, and 3% had an Express Gold card. Finally, of Americans earning less than $15,000, only 36% had a bank card, and only 1% had an Express Gold card. However, 19% of this group had gasoline cards.

10. Title _____

	Income Level		
Type of Card	*$50,000 or more*		
Bank Credit Card	*94%*		

Source: _____

Use the table on page 28 to fill in the blanks.

11. _____% of Americans earning $25,000–$34,999 have a bank credit card.

12. 19% of Americans earning less than $15,000 have a _____ credit card.

13. 3% of Americans earning _____ have an Express Gold card.

14. _____% more Americans earning $50,000 or more have a bank credit card than a gasoline credit card.

Use the table to decide if each statement is true or false.

T F 15. Of Americans earning $25,000–$34,999, 14% hold an Express Gold card.

T F 16. Americans earning $50,000 or more are the highest percent of credit card holders for every card listed.

T F 17. Less than 50% of Americans earning $25,000–$34,999 hold a bank credit card.

T F 18. Of Americans earning less than $15,000, 1% hold an Express Gold card.

Make two statements using the data in the table. The first one has been started for you.

19. The bank credit card _____

   _____

20. _____

   _____

Bar Graphs

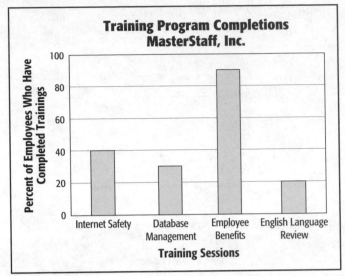

Training Program Completions
MasterStaff, Inc.

Percent of Employees Who Have Completed Trainings (vertical axis: 0, 20, 40, 60, 80, 100)

Training Sessions: Internet Safety, Database Management, Employee Benefits, English Language Review

Source: Master Staff, Inc.

Have you ever seen a graph similar to this one?

What is being compared on the graph?

How do you read a **bar graph** like the one above? Just as with a table or chart, you need to pay attention to the title and source line. Listed below are other major parts of a bar graph that will help you interpret the data.

- *The vertical axis, the horizontal axis, and their labels*

On the graph above, the **vertical axis** represents percent while the **horizontal axis** shows different training programs.

- *The bars*

Bars can be vertical (as in the graph above) or horizontal. Value is shown by the height or length of each bar. Each bar on the graph above represents a different training program; for example, Internet Safety and Database Management.

- *The scale and range*

A **scale** marks the length of each bar and is labeled in equal increasing values. The **range** is the lowest number on the scale through the highest. The scale on the graph above shows percents in increases of 20 with a range from 0 to 100.

You may already know how to read a bar graph.

EXAMPLE What percent of employees completed the internet safety training?

STEP 1 Find the bar labeled *Internet Safety* on the horizontal axis.

STEP 2 Move your finger from the top of the bar across to the corresponding value on the vertical axis. Or place the edge of a paper across the top of the bar to the vertical axis.

STEP 3 The value is 40.

ANSWER: The graph tells you that **40% of employees completed internet safety training.**

Now figure out what percent of employees completed database management training. Can you see that the *Database Management* bar rises to a point between two points labeled on the scale?

The range of percents given on the graph is from 0% to 100%, and the values are labeled in multiples of twenty (20, 40, 60, and so on).

The value for database management is halfway between 20 and 40; therefore, **30% of employees received this training.**

Take a few minutes to determine from the graph on page 30 what percent of employees completed the trainings listed.

1. Employee Benefits: _____

2. English Language Review: _____

Bar graphs are useful for more than just showing specific data. A quick look at a bar graph allows you to make comparisons very easily.

For example, did more employees complete English Language Review or Internet Safety training?

All you need to do is compare the lengths of the two bars. The internet safety bar is taller than the English language review bar. Therefore, **more employees completed internet safety training.**

3. Critical Thinking On the Job If you worked for MasterStaff, Inc., how might you use this graph? Could you use this data to make some decisions about employee training?

Now apply what you have just learned to another bar graph. Notice the similarities and differences between it and the one on page 30.

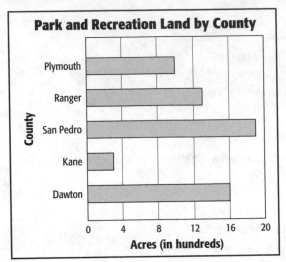

Park and Recreation Land by County

4. What is the title of the graph?
   Title: _____

5. What does the vertical axis represent?
   Vertical Axis: _____

6. What does the horizontal axis represent?
   Horizontal Axis: _____

7. What is the range of values on the scale?
   Range: _____ to _____

8. The scale is measured in multiples of _____.
   Multiples of _____

Now use the graph to make some comparison statements. Do not look at specific data points. Instead, use the relative lengths of the bars to make these statements.

9. Dawton County has more park and recreation land than

   _____

10. Kane County _____

   _____

11. _____

   _____

Use the information given below to construct a bar graph. Be sure to include a title, labels on both axes, and a source line.

(**Hint:** Label the vertical axis from 0 to 120. Use multiples of 10.)

The International Atomic Energy Agency stated that in 2010 the United States had 104 nuclear reactors in operation. The agency compared this number to the 58 reactors operating in France, the 32 operating in Russia, and the 18 operating in both Canada and Germany.

12.

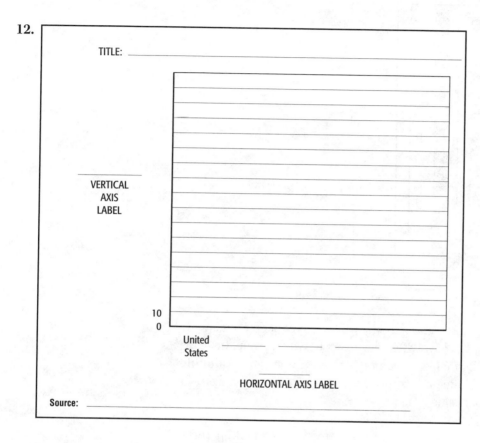

TITLE: _____

VERTICAL
AXIS
LABEL

10
0

United
States

HORIZONTAL AXIS LABEL

Source: _____

Now use the graph to make some comparison statements.

13. In 2010 the United States had _____

_____

14. Canada had _____

_____

15. France had _____

_____

Double Bar Graphs

What does the key tell you on this graph?

Why would double bars like these be used to present data?

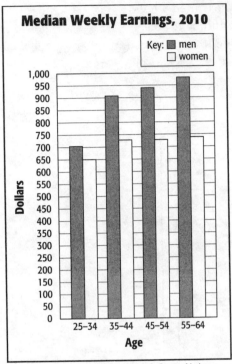

Median Weekly Earnings, 2010

Key: ■ men
□ women

Dollars (y-axis): 0 to 1,000 in increments of 50

Age (x-axis): 25–34, 35–44, 45–54, 55–64

Source: Bureau of Labor Statistics, U.S. Dept. of Labor

This bar graph gives you information about two groups of people—men and women—and uses a **key** to show which bar refers to which group. Instead of creating separate graphs for men and women, a single graph called a **double bar graph** can be used to include both sets of data. You may also have seen graphs with three or more bars side by side.

On a multiple bar graph, more data can be presented without making the graph too hard to read.

For example, if you were interested in finding information about women's median weekly earnings, what type of bar would you look at?

And if you wanted to know what the median weekly earnings were for men age 55–64, where would you look?

Did you answer the **unshaded bars** to the first question and the **last shaded bar** to the second question?

As with all bar graphs, comparisons between groups can easily be made by looking at the length of the bars. In double bar graphs, comparisons can be made *within* groups and *between* groups.

Using the graph on page 34, you can compare median weekly earnings of men vs. women in the same age group.

You can also compare earnings among different age groups.

Write a statement that compares the median weekly earnings of one age group to those of another.

Statement 1: _____

_____

Now write a statement that compares the median weekly earnings of men to those of women.

Statement 2: _____

_____

Here are some sample statements. Do yours sound like these?

- In the United States, men and women age 35–44 earned more than men and women age 25–34.

- In the United States, men of all ages earned more than women in the same age group.

Use the graph on page 34 to decide if each statement is true or false.

T F **1.** The group with the highest median weekly earnings in 2010 was men age 45–54.

T F **2.** At no time between the age of 25 and 54 were women's median weekly earnings higher than the *lowest* weekly earnings for men between the age of 25 and 54.

T F **3.** Men's median weekly earnings in 2010 were at their highest at age 55–64.

T F **4.** Women at age 25–34 earned a median weekly earnings of about $650 in 2010.

Line Graphs

How is this line graph different from a bar graph?

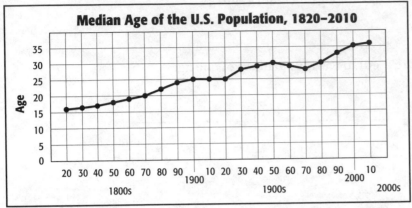

Median Age of the U.S. Population, 1820–2010

Source: U.S. Census Bureau

Like a bar graph, a **line graph** uses values plotted along vertical and horizontal axes to display data visually.

You read a line graph in much the same way that you read a bar graph—by finding the point at which the vertical axis value and the horizontal axis value meet, or intersect. Instead of looking at a bar, however, you are looking at a line that may rise or fall, depending on the data.

EXAMPLE   **What was the median age in the United States in 1920?**

STEP 1 Find the year 1920 along the horizontal axis.

STEP 2 Find the point on the line graph directly above 1920.

STEP 3 Move your finger across the graph and find the corresponding value on the vertical axis.

   What is the vertical axis value?

ANSWER: The median U.S. age in 1920 was **25 years old**.

A line graph shows values as they change over time. In the line graph on page 36, what value changes over time?

You're right if you said **median age in the United States.** In general, what changes took place between 1820 and 2010? Did the median age go up or down?

Just by looking at the angle of the line, you can see that **the median U.S. age went up,** except in the 1950s and 1960s, when it went down slightly.

Math Recap

Do you know what *median* means? In this case, it means that half the U.S. population is above the age on the graph, and half the population is below. You will learn more about median on page 54.

Use the graph on page 36 to answer the following questions.

1. What was the median U.S. age in 1950?

2. By how much did the median U.S. age go up between 1870 and 1920?

3. A woman is 39 in 2010. Is she above or below the median age in the United States?

Make three statements using the graph on page 36. The first one has been started for you.

4. The median U.S. age was the same in 1980 as it was in

_____

5. _____

_____

6. _____

_____

7. **Critical Thinking** Why do you think the median U.S. age has risen over the past two hundred years? Can you think of some reasons that Americans might be living longer?

More Line Graphs

How is this line graph different from the line graph in the last lesson?

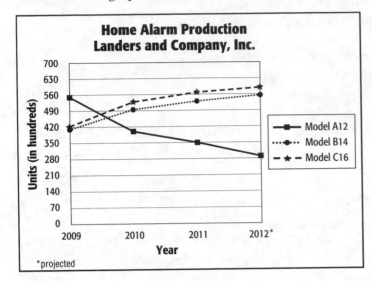

As in the line graph on page 36, this graph has years labeled on the horizontal axis, and it shows how certain values change over time. However, this line graph shows *three* separate lines, not just one. Each line represents a separate set of values. The key included on the graph shows what each line refers to.

For example, the solid line (——) refers to Alarm Model A12. What do the other two lines refer to?

(• • •) refers to _____.

(– – –) refers to _____.

You're right if you said **Model B14** and **Model C16,** in that order.

As the title and axes of the graph indicate, all three lines show the changes in home alarm production between 2009 and 2012. Each individual line refers to a particular model of alarm.

Did the production of Model A12 rise or fall between 2010 and 2011?

STEP 1 Find the line in the key that represents Model A12.

STEP 2 Find the years 2010 and 2011 on the horizontal axis.

STEP 3 Find the points on the line graph directly above 2010 and 2011.

STEP 4 Move your finger across and find the corresponding values on the vertical axis.

2010 = about 400

2011 = about 350

ANSWER: **Model A12 production fell**.

Notice that you could also answer the question above without looking at any specific values on the vertical axis. Because the data line is falling between 2010 and 2011, you know that Model A12 production fell during that time.

Use the graph on page 38 to answer the questions.

1. In general, has Model B14 production risen or fallen between 2009 and 2012?

2. Between 2010 and 2011, did production of Model C16 rise or fall?

3. Which home alarm model did Landers produce the most of in 2009?

4. Which alarm model will Landers produce the most of in 2012?

5. **Critical Thinking on the Job** Model A12 sells for a higher price to homeowners than Model B14 and Model C16. Use this information to write a statement about the production trends shown on this graph.

6. Use the data from the bar graph below to construct a line graph on a separate sheet of paper.

STEP 1 Draw and label the vertical and horizontal axes using the same values that are given on the bar graph.

STEP 2 Read the data for 2005 on the bar graph. On your line graph, put a dot or small ✕ at the same value.

STEP 3 Continue graphing data for the remaining years.

STEP 4 When you are finished, connect the data points.

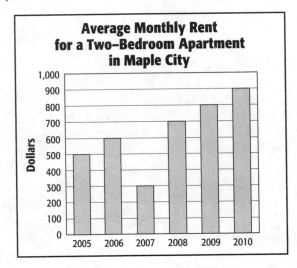

7. Which graph (the bar graph above or the line graph you have just drawn) best shows the change in rental prices?

_____

8. Write a statement comparing the highest average monthly rent with the lowest.

_____

_____

_____

9. Write a statement about Maple City rental prices between 2005 and 2010.

_____

_____

_____

Visual Statistics

Does this graph look like a line graph?

How is it different from other line graphs you have worked with?

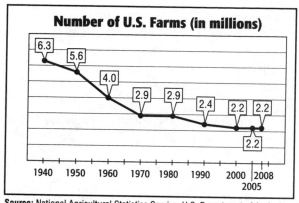

Number of U.S. Farms (in millions)

Source: National Agricultural Statistics Service, U.S. Department of Agriculture

Although the graph uses a line to show data points, it is not really a line graph. In this case, no vertical axis is shown.

Some specific points are labeled to show the number of millions of farms in 1940, 1950, 1960, 1970, 1980, 1990, 2000, 2005, and 2008. Instead of finding the exact values on a vertical axis, these points are labeled right on the graph. They're easy to read, aren't they?

Why don't all line graphs label data points like this? Here's why: Suppose you wanted to find the number of farms in 2007. It's not labeled on this graph, and there's no vertical axis with a range of values to provide additional information. On a real line graph, you could estimate the value for 2007 by using the scale on the vertical axis.

The visual statistics above are terrific if you want to know a general trend about the number of U.S. farms or if you are interested in the specific years labeled. Otherwise, a more complete line graph is better.

Use the visual statistics above to make two statements about the number of farms in the U.S. The first one has been started for you.

1. Between 1940 and 2008, the number of farms in the United States

_____

2. _____

_____

Pictographs

Why do you think data is sometimes shown on a pictograph like the one at the right?

Pictographs are helpful for making comparisons at a glance.

Pictographs are often used in magazines and newspapers because they

- have visual appeal
- are easy to read
- can be easily constructed

This pictograph shows approximations of the average time per month spent on the internet by people in selected countries.

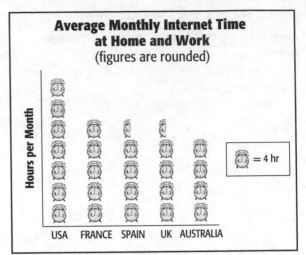

Source: Telegraph.co.uk, 09 May 2008

The key tells you that each symbol is equal to four hours of internet time.

EXAMPLE   How many hours per month, on average, does a person in the U.S. spend online?

STEP 1   Find the country on the graph.

STEP 2   Count the number of symbols next to it.

STEP 3   Multiply 7 by the value given in the key, 4.

ANSWER: On average, a person in the U.S. spends **28 hours** on the internet.

Find and compare other values found in the graph.

1. How many hours per month, on average, does a person in Spain spend on the internet?
   (**Hint:** Half of the symbol equals $\frac{1}{2}$ of 4.)

2. Of the countries shown, which country spends the least amount of time on the internet?

3. How many more hours per month, on average, does a person in France spend on the internet than a person in the UK?

4. If a person in the U.S. spends 75% of his/her internet time emailing, how many hours per month, on average, is this?

Use the chart below to construct a pictograph.

- When deciding what dollar value your symbol should represent, look at the largest number you need to graph. Be sure that this amount can be represented in 15 or fewer symbols. More than 15 symbols can become difficult for your reader to count.

- Some suggestions for symbols are circles, squares, Xs, or dollar signs. Or you can create your own symbol.

Average Salary for a Full-Time Worker JT Manning, Inc.
2005: $30,000
2006: $40,000
2007: $45,000
2008: $50,000
2009: $60,000
2010: $65,000

Write two questions you could ask about the pictograph.

EXAMPLE   Did the average salary stay the same two years in a row?

5. _____

_____

6. _____

_____

Circle Graphs

How do you think a circle graph differs from a bar or line graph?

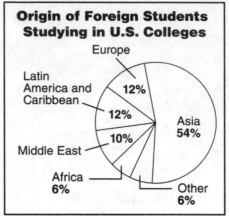

Origin of Foreign Students Studying in U.S. Colleges

Europe 12%
Latin America and Caribbean 12%
Middle East 10%
Africa 6%
Asia 54%
Other 6%

Source: Institute of International Education

A **circle graph** is a circle divided into parts. The circle represents a *whole*, and each wedge, or section, represents a *part* of that whole.

Add up the percents in the graph above. What total percent is represented by the circle graph?

The circle above represents *all*, or 100%, of the foreign students studying in U.S. colleges. Each part of that whole represents the percent from a specific area of the world.

EXAMPLE   **What percent of foreign students studying in the United States are from Africa?**

STEP 1 Find the section, or part, labeled *Africa*.

STEP 2 What percent is shown within this section?

ANSWER: 6% of foreign students studying in the United States are from Africa.

Circle graphs are very useful when you want to make comparisons. For example, just by looking at the graph above, you can see that a much larger percent of students come from Asia than from Africa or the Middle East.

Make two statements comparing the percent of students from different regions. The first one has been started for you.

1. There are fewer students from _____

_____

2. _____

_____

What does *Other* refer to on the graph? There are often groups of data that are too small to list. These are combined to form one larger section labeled *Other*. In this way, the graph still represents 100% of the data.

"Dollar" Circle Graphs

Circle graphs are usually divided into percents, as is the one on page 44. Sometimes, however, a circle graph is divided into *cents*, or parts of a dollar. Look at the graph at the right.

This graph shows you how many cents out of every dollar went toward various government functions.

For example, out of every dollar collected by the federal government, 25 cents went toward national defense. How many cents went toward grants to states and localities?

You're correct if you said **12¢.**

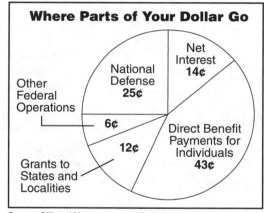

Where Parts of Your Dollar Go

Net Interest 14¢

National Defense 25¢

Other Federal Operations 6¢

12¢

Grants to States and Localities

Direct Benefit Payments for Individuals 43¢

Source: Office of Management and Budget

Make a statement comparing the amount of federal money that goes to two different government functions.

3. _____

_____

Use the information below to label the circle graph. The circle represents a total of 100¢ and the sections are proportional. For example, 43¢ is a little less than $\frac{1}{2}$ of the circle.

4. The United States Office of Management and Budget reports that 34 cents out of every dollar collected by the U.S. government comes from Social Security receipts and 43 cents out of every dollar comes from individual income taxes. Out of each dollar, it is calculated that 11 cents comes from corporation income tax. The remaining 12 cents comes from other sources.

Federal Government Dollar Where It Comes From

Source: Office of Management and Budget

Make two statements comparing where federal money comes from. The first one has been started for you.

5. More than half of federal money _____

_____

6. _____

_____

Line Plots

What was the most common number of wins by teams in the National Football Conference in 2009?

2009 Regular NFL Season Final Standings National Conference	
Team	**Number of Wins**
New York Giants	8
Philadelphia Eagles	11
Washington Redskins	4
Arizona Cardinals	10
Dallas Cowboys	11
Minnesota Vikings	12
Green Bay Packers	11
Detroit Lions	2
Chicago Bears	7
Tampa Bay Buccaneers	3
San Francisco 49ers	8
St. Louis Rams	1
New Orleans Saints	13
Atlanta Falcons	9
Carolina Panthers	8
Seattle Seahawks	5

A **line plot** is often used to show data in a more visual way than the chart above. Construct a line plot and see what advantages it has.

STEP 1 Draw a line across your paper.

STEP 2 Find the range you are working with. (In this case, the fewest number of wins is 1, and the greatest is 13.) Write these numbers under your line as shown and fill in the values in between.

STEP 3 Now use an × to record each number of wins on the number line. (The Giants had 8 wins, so put an × above the 8 on the number line.)

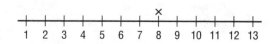

STEP 4 Continue marking the ×s in a stack above the appropriate numbers. (The values for the Eagles, the Redskins, the Cardinals, the Cowboys, and the Vikings are now added.)

What types of things can you see more clearly from a line plot than you can from a chart of data?

You may have seen a couple of things just "jump out at you" that perhaps were not as obvious when you read the chart.

- There is a **cluster** of ×s around a certain area of the line. Where is this bunch, or group, of ×s? You're right if you said **between 11 and 12.** This cluster of ×s tells you that 11 or 12 wins was a more common occurrence than, say, 2 or 3 wins.

- There are single data points on the line that are quite a bit above or below the cluster. From these, it is also easy to see the highest and lowest values in the data collected.

Line plots make it easy to compare individual values at a glance. For instance, do teams generally win about 10 games, or more than that?

Use the data below to create your own line plot.

- Read the number of representatives for each state and put an × above that value on the line plot.

- The first five states on the list have been marked for you.

1. **Number of Members in the House of Representatives, by State, 2010**

✓ Alabama	7	Louisiana	7	Ohio	18
✓ Alaska	1	Maine	2	Oklahoma	5
✓ Arizona	8	Maryland	8	Oregon	5
✓ Arkansas	4	Massachusetts	10	Pennsylvania	19
✓ California	53	Michigan	15	Rhode Island	2
Colorado	7	Minnesota	8	South Carolina	6
Connecticut	5	Mississippi	4	South Dakota	1
Delaware	1	Missouri	9	Tennessee	9
Florida	25	Montana	1	Texas	32
Georgia	13	Nebraska	3	Utah	3
Hawaii	2	Nevada	3	Vermont	1
Idaho	2	New Hampshire	2	Virginia	11
Illinois	17	New Jersey	13	Washington	9
Indiana	9	New Mexico	3	West Virginia	3
Iowa	5	New York	29	Wisconsin	8
Kansas	4	North Carolina	13	Wyoming	1
Kentucky	6	North Dakota	1		

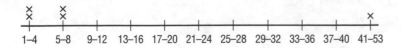

2. What is the least number of representatives from a state, and how many states have that many?

3. What is the greatest number of representatives? Are there many states that have close to this number of representatives?

4. Is there a cluster of values in a certain area of the line plot? Where is it?

5. Representation in the House is based on state population. The larger the state population, the greater the number of representatives. Make two statements about population based on the information above. The first one has been started for you.

   a. Many states have _____

   _____

   b. _____

   _____

6. **Critical Thinking** Each state in the United States is represented by two members in the Senate, regardless of the state's population. Why do you think membership in the House of Representatives is based on population? Is it fair that states with more people have more representatives?

Scatter Diagrams

An employee of a cereal manufacturer said to his boss, "I have this theory that the more sugar we put in our breakfast cereals, the more kids will like them. What do you think?" "Brilliant!" replied the boss. "Let's collect some data and check it out." Here's the data they came up with:

Cereal	Teaspoons of Sugar per Serving	Number of Children Who Liked It
Good for You	1	2
Sweeties	8	19
YumYums	4	12
SugarPlus	5	11
PlainPuffs	2	5
Sugar City	9	20
Kiddo Krunch	5	13
Sweethearts	8	17
Munchies	6	12

Is the employee's theory correct?

Can you tell by looking at the chart?

You may have noticed that it is difficult to prove the employee's theory by simply looking at the data. An excellent way to display this type of data is with a **scatter diagram.** The diagram below shows the intersection of amount of sugar and number of children.

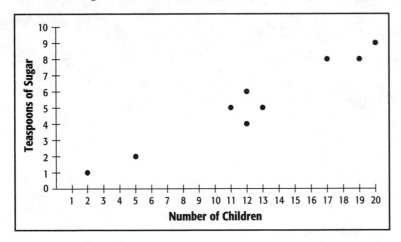

Can you tell from this diagram whether the cereal company employee's theory is true? Does more sugar mean that more kids like the cereal?

The answer is **yes.** As the number of teaspoons of sugar goes up, so does the number of kids who like the cereal. In general, as one number gets larger, the other number gets larger. This means that there is a **correlation**, or a relationship between the two values.

A research group did an informal study of the relationship between people's income and the number of years of school they had completed.

Some responses have already been plotted on the scatter diagram below.

Respondent	Years of School	Income
2E	16	$30,000
2F	10	28,000
2G	14	21,000
3A	14	20,000
3B	8	7,000
3C	9	10,000
3D	9	11,000
3E	15	30,000
3F	16	25,000
3G	16	12,000

Now plot the responses listed in the chart at the right. For each respondent, put a point at the intersection of the two values: years of school and income.

1.

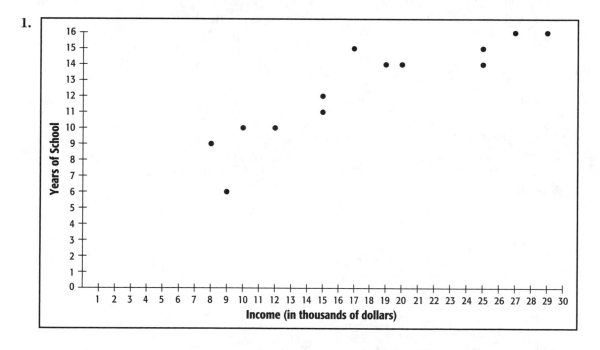

2. In general, do you see a correlation between number of school years and income? Write a statement about the data based on your scatter diagram.

_____

_____

Arithmetic Mean

"You are getting a fantastic deal on this apartment," says a landlord. "This beautiful place for only $400 per month! The average rental rate for two-bedroom apartments in this neighborhood is almost $500. If you want to live in this area, you can't do better than this place."

Sounds like a good deal, doesn't it?

Assuming that the landlord did the math correctly, could you get a two-bedroom apartment for much less than $400?

What Is a Typical Value?

You have seen how data and statistics can be represented on graphs, charts, and diagrams. Another way that statistics are often presented is in the form of **typical values.**

For example, you might say that the *most common* age of a high school senior is 17 years old. This does not mean that *all* high school seniors are 17, but that many are 17 and most are *close* to 17.

There are three different ways to describe a typical value: finding the **mean**, the **median,** and the **mode.** In fact, the landlord above is not really giving the renter enough information to make a decision about rental prices in the area. Let's see why.

Arithmetic mean, or just **mean,** is another way of saying **average**.

Follow these steps to find the mean of a set of data.

STEP 1   Add up the numbers.

STEP 2   Divide the sum by the number of values in your set of data.

Here are the rental rates in the neighborhood.

1200 Speen St.	$250	1320 Day St.	$240
1202 Speen St.	250	1322 Day St.	220
1203 Speen St.	240	1400 Day St.	250
1211 Speen St.	250	1401 Day St.	250
1220 Speen St.	250	400 North Ave.	250
320 Ashland Ave.	270	402 North Ave.	290
330 Ashland Ave.	240	404 North Ave.	270
333 Ashland Ave.	260	406 North Ave.	260
440 Ashland Ave.	230	408 North Ave.	1,700
500 Ashland Ave.	250	410 North Ave.	1,650
1111 Day St.	290	412 North Ave.	1,600
1290 Day St.	250	414 North Ave.	1,700

EXAMPLE What is the mean of the rental rates listed?

STEP 1 Add up all the rental rates in the set of data.

The total is $11,710.

STEP 2 Count the number of rental rates listed.

There are 24 rental rates in this set of data.

STEP 3 Divide the sum by the number of rental rates in the set of data.

$11,710 ÷ 24 = **$487.92**

ANSWER: The mean is **$487.92**.

As the landlord stated, the average (mean) of rental rates is "almost $500." However, look again at the rates listed. Are any of the values even close to $500? No. In fact, most of the values *center around $250.*

The mean is so high because of the four very high rental rates of $1,700, $1,600, $1,700, and $1,650. These rates are so much higher than the others that they raise the mean to an amount that is not really typical!

Is the landlord on page 52 not telling the truth? Is it possible to tell the truth but still be misleading?

Find the mean (average) in each set of data below.

1. Distance from home to workplace among employees:
   2.8 miles, 4.8 miles, 8 miles, 1.5 miles, 3 miles, 6 miles,
   2.5 miles, 3.4 miles

2. Weekly grocery bills: $90, $120, $48, $55, $37

3. Tips earned: $35, $48, $22, $18, $25, $32, $44

Median

Look at a second way to describe the rental rate data on page 52.

The **median** of a set of data is the middle value when the pieces of data are ordered from least to greatest value.

EXAMPLE 1 What is the median in this data set?

| 56 | 61 | 49 | 32 | 90 | 75 | 23 | 101 | 70 |

STEP 1 Put the set of data in order, from least to greatest value.

| 23 | 32 | 49 | 56 | 61 | 70 | 75 | 90 | 101 |

STEP 2 Select the value in the middle of the set.

| 23 | 32 | 49 | 56 | (61) | 70 | 75 | 90 | 101 |

ANSWER: The median value in this set of data is **61**.

However, when there is an even number of values in a set of data, the median is the *mean* of the middle two numbers.

EXAMPLE 2 What is the median in this data set?

20, 21, (22, 23,) 24, 25

STEP 1 Add the two middle values.

$22 + 23 = 45$

STEP 2 Divide by 2.

$45 \div 2 = \mathbf{22\frac{1}{2}}$

ANSWER: $\mathbf{22\frac{1}{2}}$ is the median of the set.

Looking back to the rental rates given on page 52, the *median* rental rate is $250.

Can you see that the *median* value of $250 is much more representative of the rental rates in this neighborhood than the mean value of $487?

Find the median in each set of data below.

1. Calculator prices among area stores: $9.50, $12.00, $14.50, $12.00, $12.50, $13.00

2. Heights of South High basketball players: 6'2", 6'10", 7'3", 7'4", 5'11"

Mode

Now look at the third way to find a typical value. The **mode** in a set of data is the value that occurs most frequently in the set.

EXAMPLE   What is the mode in this set of data?

| 21 | 15 | 70 | 22 | 21 | 71 | 21 | 15 | 40 |

How many times does each value appear?

21: 3   70: 1   71: 1

15: 2   22: 1   40: 1

ANSWER: The mode is **21** because it appears most often in the set of data.

What is the mode of the rental rates given on page 52?

1. Determine how many times each value appears in the set of data. (This process has been started for you.)

    1,700: _____2_____   260: _____

    1,650: _____   250: _____

    1,600: _____   240: _____

    290: _____   230: _____

    270: _____   220: _____

2. The value that appears most frequently is the mode. What is the mode?

    Again, in this case, the **mode** gives you a better picture of the rental rates in the area than the mean does.

    (**Note:** Sometimes there is no mode in a set of data because each value appears only once. In this case, use mean or median to express the typical value.)

Find the mode in each set of data below.

3. Prices of a 1-pound box of salt in area stores: $0.57, $0.49, $0.52, $0.49, $0.57, $0.49

4. Number of weeks on Top 40 radio: 8, 10, 5, 3, 9, 10, 10, 9, 5, 8, 4, 2

Mean, Median, and Mode

Is mode or median always a better way to show the typical value than the mean? No. In the case of the landlord on page 52, it certainly would have been fairer to the potential renter to use median or mode. However, there are other cases where the mean *does* give a more accurate picture of the typical value.

> Use the mean to show typical value *unless* there are extreme values in the set of data. Extreme values are those that are not typical of the rest of the data.

For each set of data below, find the mean, the median, and the mode. Then decide which method best represents the most typical value in each set.

1. **Dollars Collected by Charity Groups**

Cancer Fund	$1,100	Mean: _____
Grand Foundation	300	
Women for Women	1,100	Median: _____
John Kramer Fund	650	
Heart & Lung Assn.	500	Mode: _____
Dalmore Committee	230	

    Best representation of typical value: _____

2. **Attendance at Plainville Pirates Baseball Games**

June 15 game	230 people	Mean: _____
June 16 game	175 people	
June 30 game	225 people	Median: _____
July 4 game	700 people	
July 5 game	225 people	Mode: _____
July 12 game	185 people	
July 14 game	200 people	
July 20 game	180 people	
July 27 game	165 people	

    Best representation of typical value: _____

Critical Thinking with Data

Suppose a local school system is deciding how much money to allot to its high school sports program. Part of its decision is based on how much the surrounding school systems spend on their programs.

Use the graph below and your knowledge of typical values to answer the questions that follow.

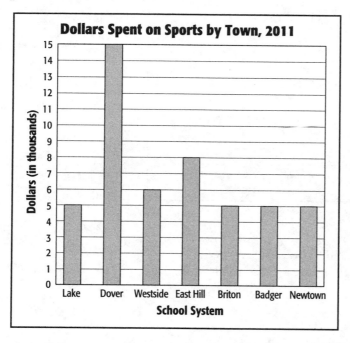

1. Suppose you are a high school football coach in the Westside school system. You want your school system to spend as much money as possible on the sports program. Would you use mean, median, or mode to show that Westside spends too little? Write a paragraph persuading the school system to spend more money. Use any data you need from the graph above.

2. Now imagine that you are a Westside school department member, and you are tired of having so much money spent on sports. You would rather have more money allotted for teachers' salaries. Would you use mean, median, or mode to show that Westside spends too much on sports? Write a paragraph defending your point of view. Use any data you need from the graph above.

ANALYZING DATA

The Purpose of a Graph or Chart

What is the purpose of the graph at the right?

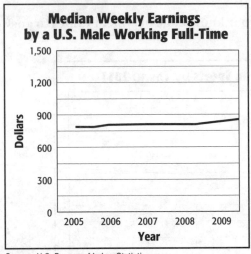

Median Weekly Earnings by a U.S. Male Working Full-Time

Source: U.S. Bureau of Labor Statistics

Your first step in analyzing data is to understand the *topic* of the data. Many times, mistakes are made in interpreting data because a reader did not understand the goal, or purpose, of a graph or chart. For example, circle the letter of the statement that accurately summarizes the purpose of the graph.

For the period 2005 to 2009, the graph shows the rise in

   a. median weekly earnings in the United States

   b. median weekly earnings for a U.S. male working full-time

   c. median weekly earnings for a U.S. male

You're correct if you chose statement **b.** What's wrong with the other two?

1. If the first statement were true, you could expect to find information about *female* workers. The graph title clearly states *male* workers.

2. If the third statement were true, you could expect to find information about *part-time* workers' earnings. The graph title specifies *full-time*.

Only statement **b** accurately summarizes the graph's purpose.

To understand the purpose of a chart, graph, or table, you need to use information from the *title* and *labels*. Ask yourself these basic questions:

- *What* data is being presented?

- *When,* or over what period of time, does the data apply?

- *Where,* or what location, is the data concerned with?

- *How* is the data presented (in dollars? in miles per hour? in average cost?)

Fill in the blanks to summarize the purpose of the chart below.

Highest-Paying Jobs* in N. Essex County, 2011	
Profession	**Average Weekly Salary**
Lawyer	$990
Airline Pilot	807
Chemical Engineer	807
Electrical Engineer	803
Aerospace Engineer	801
Physician	792
Mechanical Engineer	766

*Among fields employing at least 5,000 people

The chart shows the

_____
what?

in _____
where?

in _____
when?

in _____
how?

(among fields employing at least 5,000 people).

An accurate summary of the purpose for the chart above is:

The chart shows the _____highest-paying jobs_____
what?

in _____N. Essex County_____
where?

in _____2011_____ in _____average weekly salary_____
when?   how?

(among fields employing at least 5,000 people).

Of course, there are other ways to summarize a graph or chart. Using your own words, state the purpose of the graph. Be sure to include the *what, when, where,* and *how* information in some form.

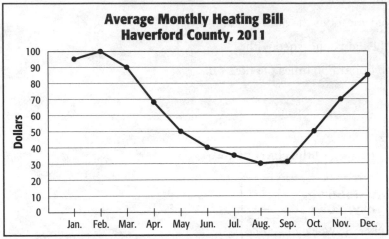

Source: County Utilities Commission

The graph shows the _____
                                          what?

in _____ in _____
        where?   when?

in _____.
          how?

Does your summary sound something like this?

The graph shows the average monthly heating bill in Haverford County in 2011 in dollars.

The Source of Data

All good graphs, charts, and tables name the **source** that collected and organized the data. A **source line** is usually printed somewhere beneath the data.

The source of any data is important to know in case you have questions about the data or its reliability.

What is the source of the data for the graph above?

_____

**Decide if the summary given for each set of data below is accurate.
If it is not, which information is inaccurate—the *what, where, when,*
or *how*?**

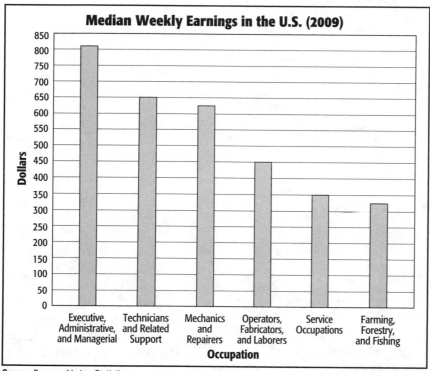

Median Weekly Earnings in the U.S. (2009)

Source: Bureau of Labor Statistics

1. The purpose of the graph above is to show the 2009 median weekly
   earnings for the best-paying jobs in the United States, in dollars.

   Is the summary accurate?

   If not, explain why.

2. The purpose of the graph at the right is to show the
   price of milk from 2008 to 2012 in dollars per quart.

   Is the summary accurate?

   If not, explain why.

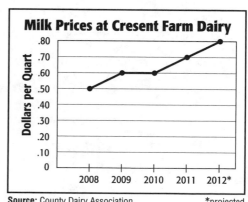

Milk Prices at Cresent Farm Dairy

Source: County Dairy Association   *projected

Americans Believe It's All Right to Lie	% Men	% Women
to protect oneself	63	52
to avoid personal embarrassment	56	48
to keep one's job	56	35
to gain a small amount of money	25	15

Source: *Press Mentor* Survey

3. The purpose of the chart above is to show how many Americans lie at certain times.

   Is the summary accurate?

   If not, explain why.

4. The purpose of the graph to the right is to show what fraction of Oldtown residents were in favor of, not in favor of, or undecided about firing the police chief.

   Is the summary accurate?

   If not, explain why.

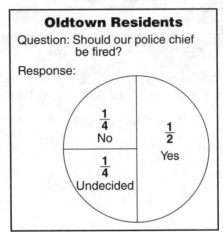

Oldtown Residents

Question: Should our police chief be fired?

Response:

$\frac{1}{4}$ No

$\frac{1}{2}$ Yes

$\frac{1}{4}$ Undecided

Source: *Free Daily Press* Survey

Reading Data: Be Specific

The vertical axis value of the data point circled is between 4 and 6. The horizontal axis value is between 2016 and 2017.

But does the graph tell you more about the circled data point?

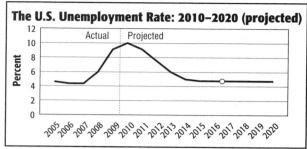

The U.S. Unemployment Rate: 2010–2020 (projected)

Source: Congressional Budget Office

In order to be as accurate as possible when making statements about specific data points, you need to use *all* of the information provided, including labels on *both* axes.

Look at the circled data point on the graph. Circle the *most accurate* and *specific* statement about that data point.

- In 2016, the U.S. had an unemployment rate of 5.
- Between 2010 and 2020, the United States unemployment rate will be 5%.
- In 2016 and 2017, the U.S. unemployment rate is projected to be 5%.

Now look at each statement in detail.

1. *In 2016, the U.S. had an unemployment rate of 5.* False

This statement does not take into account the label on the vertical axis—*Percent*. Five *percent* is the projected unemployment rate.

2. *Between 2010 and 2020, the United States unemployment rate will be 5%.* Not Specific

Not all of the years showed this 5% figure. You can figure out the specific year shown by looking at the horizontal axis. The circled data point is between 2016 and 2017.

3. *In 2016 and 2017, the U.S. unemployment rate is projected to be 5%.* Accurate

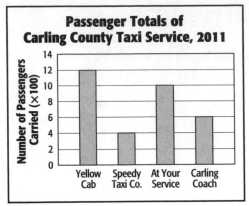

**Passenger Totals of
Carling County Taxi Service, 2011**

Source: County Taxi Association

**Tell why each of the following is *not* a precise statement about the data
in the graph above. In other words, what additional information could be
given to make each statement true or more accurate?**

EXAMPLE   In 2011, the Yellow Cab Company carried 800 times as many
passengers as Speedy Taxi Co. did.

The Yellow Cab company carried 800 more

passengers, not 800 times as many.

1. In 2011, the Yellow Cab company in Carling County carried 12 passengers.

   _____

   _____

2. At Your Service carried 1,000 passengers.

   _____

   _____

3. The Carling Coach bar goes up to 600 on the graph.

   _____

   _____

Using the graph on page 64, choose four phrases from the list below to make four different statements about the data. Be sure to use

- information from the title of the graph
- information from *both* axes
- labels

Phrases

$\frac{1}{3}$ the number of passengers

twice as many

half as many

the greatest number of

more passengers than

800 more passengers than

200 fewer passengers than

the fewest

EXAMPLE   In 2011, Speedy Taxi Co. carried the fewest passengers among the four taxi services listed.

4. _____

_____

5. _____

_____

6. _____

_____

7. _____

_____

Reading Data: Using Only the Information Given

In the United States, what state had the greatest number of hazardous waste sites in 2008? Can you tell from this graph?

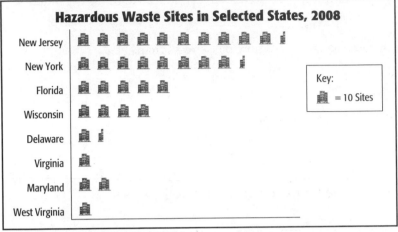

Source: U.S. Environmental Protection Agency

You might be tempted to answer New Jersey to the question above. Of the states *listed on this graph*, New Jersey does, in fact, have the greatest number. However, look closely at the title of the graph. Notice that only *selected* states are shown. A state that is not shown might have a greater number of hazardous waste sites than New Jersey. *The graph simply does not give us that information.*

In this lesson, you will work on using *only* the information on the graph.

Make a statement about Wisconsin's hazardous waste sites using the data from the graph. Remember to include only information from
- **the title**
- **the data given**
- **the key**

Statement: _____

_____

_____

Does your statement sound something like this?

Wisconsin has about 40 hazardous waste sites.

The following statements about the graph *are not accurate*. Put a check mark next to the information listed below each statement that is *not* supported by the graph and therefore *cannot* support such a statement.

1. According to the U.S. Environmental Protection Agency, Florida residents were more willing to store hazardous waste than the citizens of Virginia.

   _____ **a.** There are more hazardous waste sites in Florida than in Virginia.

   _____ **b.** Florida residents are willing to store hazardous waste.

   _____ **c.** Virginia residents fought hard against hazardous waste sites.

2. Maryland, with 20 hazardous waste sites, is a more dangerous place to live than West Virginia, with only ten sites.

   _____ **a.** The sites in Maryland are better contained than those in West Virginia.

   _____ **b.** Maryland has more sites than West Virginia.

   _____ **c.** Maryland also has more crime than West Virginia.

3. West Virginia has the fewest hazardous waste sites in the United States.

   _____ **a.** West Virginia has ten hazardous waste sites.

   _____ **b.** Hazardous waste sites are listed for all 50 states.

   You should have checked the following information that cannot be supported by the graph.

1. **b., c.** From the given information, you have no way of knowing how residents reacted.

2. **a., c.** No information is provided about the condition of the sites or the crime rate for any of the states.

3. **b.** Hazardous waste sites are listed for only eight states.

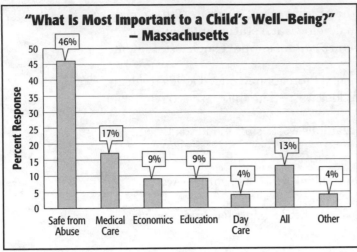

"What Is Most Important to a Child's Well-Being?"
– Massachusetts

Source: *Boston Parents' Paper*

State why each of the following is *not* an accurate statement about the data in the graph above.

<table>
<tr><td>EXAMPLE</td><td>Nine percent of people responding to a Massachusetts poll felt that their children's education was unsatisfactory.</td></tr>
</table>

The graph refers to how people answered the question, "What is most important to a child's well-being?" Respondents were not necessarily dissatisfied.

1. Just under half of all Massachusetts children were victims of abuse.

_____

_____

2. Forty-six people responded that safety from abuse was the most important factor to a child's well-being.

_____

_____

3. More Americans believe that medical care is more important to a child's well-being than quality of day care services.

_____

_____

The Language of Estimation

What is the domestic crude oil production expected to be in 2025?

Is it easy to determine from this graph? Why?

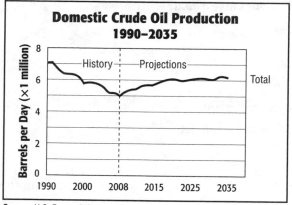

Domestic Crude Oil Production 1990–2035

Source: U.S. Energy Information Administration

Many graphs are not meant to display exact figures. For example, on the graph above, it is impossible to tell the *exact* number of barrels in any year. All you can really see is that the data point for 2025, for example, is somewhere around 6.

This doesn't mean that the graph is poorly constructed. Different graphs have different purposes. If you want to find a value on a graph like the one above, you'll need to **estimate.**

There are many ways to express estimates. Use your knowledge of the words in italic to decide if each statement is true or false according to the graph above. Circle T or F.

T F **1.** The number of barrels of oil produced in the U.S. *declined* between 2000 and 2008.

T F **2.** The amount of U.S. crude oil production in 2025 is projected to be *about* 6 million barrels per day.

T F **3.** The *approximate* U.S. crude oil production in 2000 was 5 million barrels per day.

You are correct if you found that the first two statements are true. Did you find that the third statement is false? The data point is clearly close to the 6 million line, not the 5 million line. Therefore, it is more accurate to say that the approximate production was 6 million.

There are many ways in which estimates can be expressed. Look at the following list of words and phrases taken from a newspaper. All of them tell the reader that the number given is an estimate.

- *almost* a quarter of a million dollars
- the lot measured *just over* 13 acres
- totaling *about* a thousand
- *less than* the average of $14,500
- needing *close to* 7 hours to complete

Use the graph to fill in the blanks below. Choose your answers from the list given.

Choose from this list:

less than
more than
almost
approximately
estimated to be
between

(**Note:** More than one choice may be correct.)

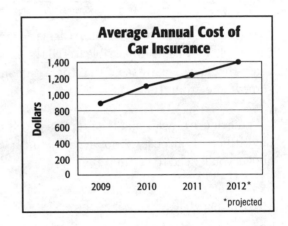

Average Annual Cost of Car Insurance

1. In 2010 the average annual cost of car insurance was
   _____ $1,000.

2. The average annual cost of car insurance was _____
   $800 and $1,000 in 2009.

3. The difference in annual car insurance between 2011 and 2012 is
   expected to be _____ $200.

Make two statements using estimates of the data on the graph.

4. _____

   _____

5. _____

   _____

Accurate Estimating

Read the two newspaper articles below.
Do they contradict each other?

The Daily Journal

...in fact, in the year 2020, the state of Florida will need close to 12,000 family physicians, and Arizona will need about a quarter that many....

Daily Express

...the projection for the number of physicans needed in Florida in 2020 is only about 11,000—less than twice the number the state had in 2006....

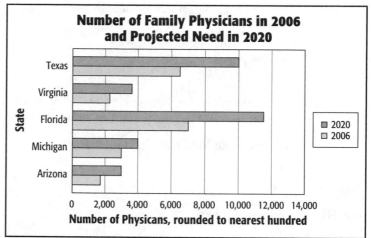

Source: AAFP Workforce Report, 2006

Which newspaper is reporting the correct figures? Will Florida need 11,000 or 12,000 family physicians in 2020?

Remember, *estimates are not exact*. The fact is that *both* newspaper articles above give acceptable estimates of the data.

Estimate an answer to the problem at the right. Do *not* find an exact number.

$$3{,}675 + 199 \approx$$

Which did you find as your estimate: 3,875; 3,900; or 4,000?

Whichever estimate you found, you are correct. In fact, any figure close to one of those above is a good estimate!

Depending on how close to the actual answer you need to be, an estimate can have a wide range of values.

Why Estimate?

Why are estimates used so often? Here are some reasons.

1. **Estimates are easier to calculate with.**

    Circle **a** or **b** to show which calculation is easier.

    a. What is 40% of 1,000 people?

    b. What is 39.8% of 1,019 people?

2. **Estimates are easier to read.** A number such as 998,786 is awkward to read and say. The estimate of one million is more easily understood.

    Say the following numbers out loud and circle the letter of the number that is easier to read.

    a. 2,000,000

    b. 2,001,112

3. **Exact figures can change frequently.** The exact number of people living in Minneapolis, for example, changes every day. Estimates are used because they are as close as we can come to the actual number.

    Which of the following questions needs only an estimate for an answer?

    a. How many people are employed in the auto industry?

    b. How many passengers can the new Ford van carry?

4. **Exact figures are often not necessary.**

    As a consumer, which of the following would you want an exact figure for?

    a. the total cost of a microwave oven you want to buy

    b. the number of microwave ovens a store has in stock

Did you circle **a** for each question above? If so, you are understanding the purpose of estimation!

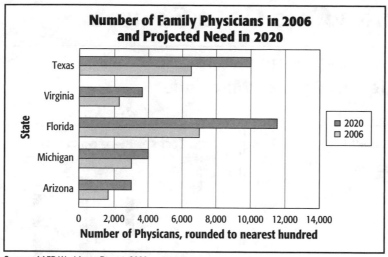

Number of Family Physicians in 2006 and Projected Need in 2020

Source: AAFP Workforce Report, 2006

Use the data to fill in the blanks that follow. Be sure to use words such as *about* or *approximately* when appropriate.

1. In 2020 Virginia will need _____ family physicians.

2. In 2006 Texas had _____ times as many family physicians as Michigan had that year.

3. _____ and _____ will need about the same number of family physicians in 2020.

4. Of the states listed, _____ will need the largest increase in family physicians between 2006 and 2020.

Look in a newspaper or magazine and find two places in which an estimate is used instead of an exact number.

5. How do you know an estimate is being used?

    _____

    _____

6. Why do you think an estimate is used instead of an exact number?

    _____

    _____

Using More than One Data Source

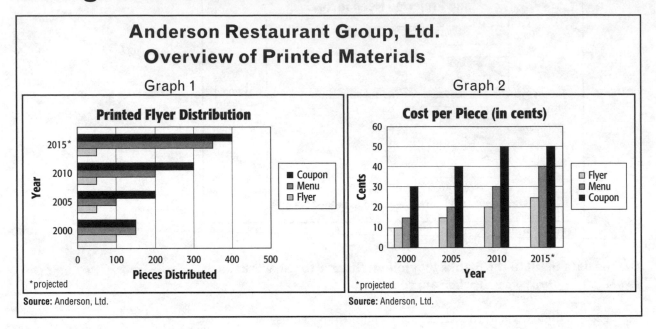

Anderson Restaurant Group, Ltd.
Overview of Printed Materials

Graph 1

Printed Flyer Distribution

Graph 2

Cost per Piece (in cents)

Source: Anderson, Ltd.

Source: Anderson, Ltd.

What is the purpose of Graph 1?

What is the purpose of Graph 2?

What kinds of things can you learn from looking
at both graphs together?

**Complete the following statements using your data summary skills
and filling in the what, where, when, and how information.**

Graph 1 above shows _____

what?

_____   _____

by?   when?

according to Anderson, Ltd.

Graph 2 shows _____

what?

_____   _____

by?   when?

according to Anderson, Ltd.

Your summaries should sound something like this:

Graph 1 shows the **number of menus, coupons, and flyers delivered
by Anderson Restaurant Group between 2000 and 2015, according to
Anderson, Ltd.**

Graph 2 shows **the cost per menu, coupon, and flyer, in cents,** according to Anderson, Ltd.

What information can you get by using *both* graphs that you can't get by using just one of them?

For example, the following question can be answered only if you use *both* graphs.

EXAMPLE   How much did it cost to print the coupons distributed by Anderson in 2010?

STEP 1   Graph 1 tells you that in 2010, Anderson distributed 300 coupons.

STEP 2   Graph 2 tells you that in 2010, each coupon cost 50 cents.

STEP 3   What is 300 times 50 cents?

    $300 \times 50 = $ **$150.00**

ANSWER: It cost Anderson **$150** to print coupons in 2010.

**Which of the graphs on page 74 do you need to answer these questions?
Circle Graph 1, Graph 2, or Both.**

1. How many coupons did Anderson   Graph 1   Graph 2   Both
    distribute in 2005 and 2010 combined?

2. How much more is Anderson   Graph 1   Graph 2   Both
    projected to spend on menu printing
    than on flyer printing in 2015?

3. Did the amount spent on coupon   Graph 1   Graph 2   Both
    printing in 2005 exceed the amount
    spent on flyers in 2010?

4. In 2015, how much more will it cost to   Graph 1   Graph 2   Both
    print a menu than a flyer?

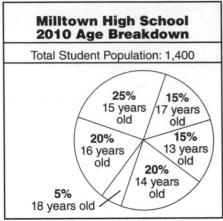

**Milltown High School
2010 Age Breakdown**

Total Student Population: 1,400

25% 15 years old
15% 17 years old
20% 16 years old
15% 13 years old
20% 14 years old
5% 18 years old

Source: Milltown School Committee

Students Participating in School Activities	
Activity	**Number of Participants**
Student government	210
Music/Band	140
Intramural sports	425
Library club	80
Social committee	120

For part a of each question, decide whether you need to use the graph, the chart, or both to answer the question. If the question cannot be answered using any of the information above, circle Neither.

For part b, answer the question if possible.

5. What fraction of the Milltown student population participated in music/band activities in 2010?

   a. Graph   Chart   Both   Neither

   b. Answer:

6. The age breakdown for student government was the same as for the entire school. How many 16-year-olds were in student government?

   a. Graph   Chart   Both   Neither

   b. Answer:

7. How many 13-year-olds attended Milltown High?

   a. Graph   Chart   Both   Neither

   b. Answer:

8. How many girls played intramural sports at Milltown High?

   a. Graph   Chart   Both   Neither

   b. Answer:

Putting Data in Different Forms

What percent of Johnson Public Housing residents were registered to vote in 2008?

Can you answer this question based on the information given?

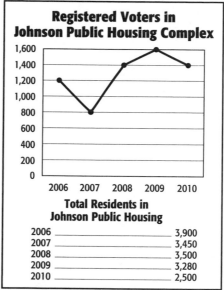

Registered Voters in Johnson Public Housing Complex

Total Residents in Johnson Public Housing

2006	3,900
2007	3,450
2008	3,500
2009	3,280
2010	2,500

Source: League of Women Voters

The group of data above gives you *numbers*, not *percents*. How can you change the form of the data to answer the question?

EXAMPLE 1 **Find the percent of residents registered to vote in 2008.**

STEP 1 Find the total number of residents in 2008.

There were 3,500 residents. (the *whole*)

STEP 2 Estimate how many people in the complex were registered to vote that year.

There were about 1,400 registered voters. (the *part*)

STEP 3 Divide the *part* by the *whole*.

1,400 ÷ 3,500 = 0.40 or **40%**

ANSWER: 40% of Johnson Public Housing residents were registered voters in 2008.

Math Recap
To find out what percent one number is of another number, divide the *part* by the *whole*.

EXAMPLE 2 Use the data shown on page 77 to find what percent of Johnson Public Housing residents were registered to vote in 2007.

STEP 1 How many total residents are there?

3,450

STEP 2 How many registered voters are there?

800

STEP 3 Divide the part by the whole.

$800 \div 3,450 = 0.2319$

STEP 4 Change the decimal to a percent.

$0.2319 = $ **23.19%**

ANSWER: About 23% of Johnson Public Housing residents were registered to vote in 2007.

Why Change the Form of Data?

Why is it useful to see the data on page 77 changed to percent? The following activity will help you see why.

Compare the **number** of registered voters in 2009 to those in 2010. Did the number go up or down?

The number of registered voters in the Johnson Public Housing Complex **fell** from 1,600 in 2009 to 1,400 in 2010.

Why do you think this drop occurred? Put a check mark next to any reasons that seem sensible to you.

fewer volunteers working to register people

no appealing candidates for office in 2010

apathy or resentment among residents

other: _____

1. Use the data on page 77 to figure out the *percents* of registered voters in the box below. Then plot the voters on the graph below.

   (**Remember:** to find percent, divide the part (number of registered voters) by the whole (number of residents). Percents for 2007 and 2008 have been computed for you.)

2006:	1,200 ÷ 3,900 is about _____ %
2007:	800 ÷ 3,450 is about 0.23, or 23%
2008:	1,400 ÷ 3,500 is about 0.40, or 40%
2009:	1,600 ÷ 3,280 is about _____ %
2010:	1,400 ÷ 2,500 is _____ %

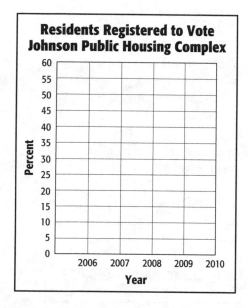

2. **a.** What happened to the percent of registered voters between 2009 and 2010? Did the percent rise or fall?

   b. Why do you think this happened?

   Perhaps you already figured out that the number of registered voters fell because the total number of residents fell as well. But the percent of registered voters was greater than any other year!

   Changing the form of data can help you see things that might not be obvious at first.

Here's another example of putting data into different forms.

Territory Sales Figures, 2011		
Sales Representative	Region	Sales (in dollars)
David Burns	Northeast	$475,000
John Fay	South	290,000
Lisa Gagnon	Northeast	325,000
Elijah Collier	Northeast	190,000
Tamara Jenkins	South	315,000
Julio Kim	Midwest	219,000
Juanita Nalda	Midwest	220,000
Laura Rice	South	110,000
Rena Sanchez	South	328,000
Herm Thomas	Midwest	200,000
Raul Whitney	Northeast	150,000

Suppose you need to calculate a company's total dollars in sales for each region in 2011. The chart above contains the only data you have available to you. The chart is organized alphabetically by sales rep. Can you calculate total regional sales?

EXAMPLE Transform the data into a table that best summarizes the regional sales totals. A table has been started for you.

STEP 1 List the different regions in either a column or a row.

Northeast	
South	
Midwest	

or

Northeast	South	Midwest
1,140,000	1,043,000	639,000

STEP 2 Add up the sales figures for each region.

Northeast: (475 + 325 + 190 + 150) × 1,000 = **1,140,000**

South: (290 + 315 + 110 + 328) × 1,000 = **1,043,000**

Midwest: (219 + 220 + 200) × 1,000 = **639,000**

STEP 3 Finish writing the totals in the appropriate column or row.

Although the original purpose of a set of data might be different from your purpose, changing the *form* of the data can often help you get the information you need.

Change the form of the data by following the steps indicated.

3. Use the steps below to find what *percent* of patients have each different insurance plan. Your information source is the graph below.

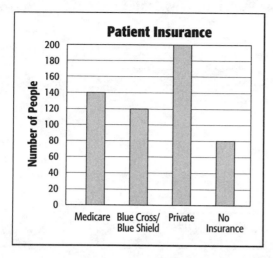

STEP 1 Find the total number of patients by adding up the numbers from each insurance category.

STEP 2 To find each percent, divide the *part* (each insurance category) by the *whole* (total patients).

STEP 3 Fill in the chart started below.

Insurance	Medicare	Blue Cross/ Blue Shield	Private	No Insurance
Percent of Patients				

4. An agency estimates that the total number of homeless people in Sun County is 9,500. Use this and the information in the chart to find the *number* of homeless people that were counted in each area, and add the information to the chart (0.37 × 9,500, and so on).

Sun County Homeless Population

Downtown Area	37% or _____
Central Square	29% or _____
East Side	22% or _____
West Side	10% or _____
Outlying	2% or _____

Seeing Trends/Making Predictions

What is a trend?

Do you see a trend on the graph at the right?

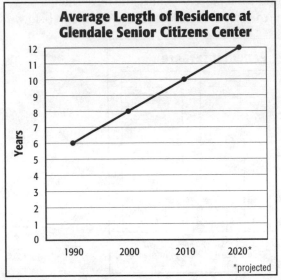

Average Length of Residence at Glendale Senior Citizens Center

Years

12 · 11 · 10 · 9 · 8 · 7 · 6 · 5 · 4 · 3 · 2 · 1 · 0

1990   2000   2010   2020*

*projected

Source: Senior Citizens Network

A **trend** is a general movement following a certain pattern or course. To understand what a trend is, fill in the blanks.

- Between 1990 and 2000, the average length of residence at Glendale Senior Citizens Center rose by _____ years.

- Between 2000 and 2010, the average length rose by _____ years.

- Between 2010 and 2020, the average length is expected to rise by _____ years.

Did you notice something repeating? Every 10 years, the average length of residence goes up by 2 years. The graph above indicates a pattern, or *trend*, in average length of residence at Glendale Senior Citizens Center.

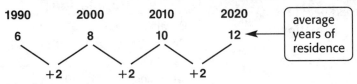

1990   **2000**   **2010**   **2020**

6   8   10   12   ← average years of residence

+2   +2   +2

If the trend continues, what will the average length of residence be in the year 2030?

To answer the question, follow the trend by adding 2 years to the average length of residence in 2020.

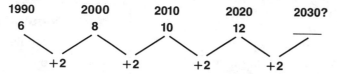

The average length of residence in the year 2030 will be **14** (12 + 2), *if the trend continues.*

General Trends

The trend on the graph on page 82 showed a rise of exactly the same number of years every 10 years (or every decade). Sometimes trends are more general—they follow a pattern, but one that is not as simple as the trend you saw before. Look at an example.

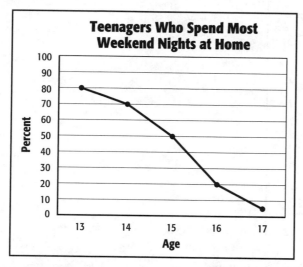

Do you see a trend concerning teenagers who stay home most weekend nights?

As you can see, there is certainly a *downward* trend with increasing age, but the size of the drop is different at each jump in age.

If the trend continues, will the percent of 18-year-olds who stay at home most weekend nights be *higher* or *lower* than the percent of 17-year-olds?

You're right if you said **lower.** Although you cannot tell how much lower, you can see from the *general trend* of the data that the percent will go down.

Projected Data

Have you ever seen numbers in a newspaper or work report that are labeled **projected?** This label means that the data is not yet available. Instead, experts are making an estimate based on a trend. The chart shows a *projection* of the world's population in 2025. The estimate is based on population totals measured in previous decades.

World Population	
Year	Population
2000	6,085,000,000
2010	7,255,000,000
2025*	8,000,000,000

*projected
Source: U.S. Census Bureau

Use the data below to solve problems 1 and 2.

Approximate Monthly Rent for a Two-Bedroom/Two-Bathroom Apartment in Del County

Key: $ = $100

1. If the trend shown continues, approximately how much would the monthly rent for a two-bedroom/two-bathroom apartment be in 2010 in Del County?

2. Make a statement about the approximate monthly rent in 2011.

_____

_____

Use the list of phrases below to write four statements about the data displayed on the graph.

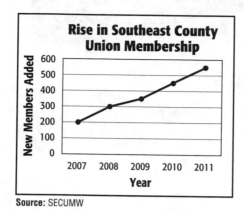

Rise in Southeast County Union Membership

Source: SECUMW

if the trend continues
rose steadily
about 50 fewer

was less in 2010
approximately 100 more
is projected to be

EXAMPLE   *The number of members added in 2011 was expected to be greater than those added in any prior year beginning in 2007.*

3. _____

_____

4. _____

_____

5. _____

_____

6. _____

_____

What Are the Chances?

"I'll probably get put on the late shift at work," said Dan, a new employee. "Personnel told me that 9 out of 10 new people start out there."

Why does Dan think he'll be put on the late shift?

Is Dan *sure* what shift he'll be put on?

Probability is the chance that an event will occur. To picture probability, use coin tossing. Toss a coin in the air and let it fall. Is one side more likely to be faceup?

1. Toss a coin 10 times and record the number of heads and tails.

   First 10 tosses:

   Heads _____

   Tails _____

2. Now toss the coin 10 more times and again record the results.

   Second 10 tosses:

   Heads _____

3. Write a statement about the total number of heads vs. the total number of tails.

   Tails _____

   _____

   _____

You can show the results of the coin-tossing experiment as a fraction or as a ratio.

$$\frac{\text{number of times outcome occurred}}{\text{total number of outcomes}}$$

Using the information from above, write fractions showing the number of tails and heads in *20* tosses.

$\dfrac{\boxed{}}{20} \quad \begin{array}{l}\text{number of tails}\\ \text{total tosses}\end{array}$ $\qquad$ $\dfrac{\boxed{}}{20} \quad \begin{array}{l}\text{number of heads}\\ \text{total tosses}\end{array}$

Fractions or ratios such as these can be used to *estimate the probability* of tossing either heads or tails. In fact, the more times you toss the coin and record the outcome, the closer you will come to knowing the actual probability. By now, you may realize that the chance of tossing heads (or tails) is close to $\frac{1}{2}$.

Here's how to represent probability.

$$\text{Probability of an event} = \frac{\text{number of desired outcomes}}{\text{total number of possible outcomes}}$$

In the case of the probability of tossing heads, the desired outcome is heads. Therefore, the probability of tossing heads is expressed as

$$\text{probability} = \frac{1}{2} \quad \begin{matrix} \leftarrow \text{(because there is 1 head on the coin)} \\ \leftarrow \text{(because there are 2 possible outcomes, heads or tails)} \end{matrix}$$

Go back and compare this fraction to the other fractions you have worked with in this experiment. What do you notice?

When you are dealing with coin tossing, all fractions and probabilities are very close to $\frac{1}{2}$.

You have learned that probability depends on the number of *desired outcomes* and the total number of *possible outcomes*.

When you roll a number cube, the chances are the same that any one of the numbers 1 to 6 will appear faceup.

EXAMPLE   What is the probability of rolling a 6?

$$\frac{1}{6} \quad \begin{matrix} \leftarrow \text{(there is one 6 on the number cube)} \\ \leftarrow \text{(number of sides on the number cube)} \end{matrix}$$

4. What is the probability of rolling a 1?

5. What is the probability of rolling an odd number?

6. What is the probability of rolling a number divisible by 3?

Now let's return to Dan's situation from the beginning of this lesson.

7. Write a fraction representing the probability that Dan will be put on the late shift.

8. What is the probability that Dan will *not* be put on the late shift?

Probabilities Between 1 and 0

What is the probability that you will fly to
the moon tomorrow?

Express this probability as a fraction.

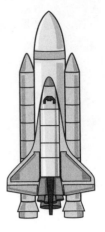

$\dfrac{0}{1}$ ← number of outcomes
← number of possible outcomes
(There's only one possibility: you won't go.)

The chances that I will fly to the moon tomorrow are 0.

What is the probability that tomorrow will follow today?
$\dfrac{1}{1}$ ← number of outcomes
← number of possible outcomes
(Tomorrow always follows today.)

It is virtually a sure thing that tomorrow will follow today. Therefore,
you can express this probability as 1.

- A probability of 1: The event will certainly happen.

- A probability of 0: It's not possible for the event to occur.

- All other probabilities are expressed as fractions between 0 and 1,
  just as you learned in the last lesson.

**Imagine that the numbered cards below are facedown on a table.
You pick up one card, not knowing which one you are choosing.**

$$\boxed{2}\ \boxed{4}\ \boxed{6}\ \boxed{8}$$

1. What is the probability that the number on the
   card you chose is a 4?

2. What is the probability that the number on the
   card is greater than 3?

3. What is the probability that the number on the
   card is divisible by 2?

4. What is the probability that the number on the
   card is greater than 8?

Look at each of the following situations and circle whether it has a probability of 0 (no chance of occurring), 1 (a sure thing), or somewhere in between.

What are the chances that

5. you will eat a hamburger in the next year?

    0   in between   1

6. it will rain somewhere in the world today?

    0   in between   1

7. someone in your state will die tomorrow?

    0   in between   1

8. you will win a million dollars in the lottery?

    0   in between   1

9. a plane is taking off now?

    0   in between   1

10. we will have world peace tomorrow?

    0   in between   1

11. a woman is giving birth right this minute?

    0   in between   1

12. the sun is shining somewhere in the world?

    0   in between   1

13. everyone will be using public transportation by the year 2020?

    0   in between   1

14. the Cubs will win the World Series?

    0   in between   1

15. Write down three events that have a probability of 0. Use your imagination!

16. Write down three events that have a probability of 1. Be creative.

Using Data to Estimate Probability

Suppose you work in a Social Security office. In June 2010, a person who receives Social Security benefits approaches your desk.

What are the approximate chances that this person is a retiree?

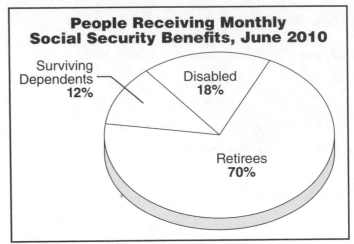

Source: Social Security Administration

Look at the data in the circle graph. The chances that a person who receives Social Security benefits is retired are pretty good—a little more than $\frac{3}{5}$ or 3 out of 5. How do you know this?

70% of people receiving benefits are retired.

$\frac{70}{100}$ ← percent of retired people
← total percent of people receiving benefit

> **Math Recap**
> Remember that 70% means 70 out of 100.

This relationship can also be expressed in terms of a probability.

$\frac{70}{100}$ is equal to $\frac{7}{10}$.

Out of every 10 people receiving Social Security benefits, approximately 7 are retired.

Therefore, you can say that the chances are about 7 out of 10 that the person who approached your desk is a retiree.

What are the chances that the person who approached your desk is *not* a retiree?

1. According to the graph, what percent of people receiving benefits are *not* retired?

   $100\% - 70\% = $ _____

2. Express this percent as a fraction, estimate, and reduce.

   $$\frac{\square}{100} \approx \underset{\text{estimate}}{\underline{\hspace{2cm}}} = \underset{\text{reduced}}{\underline{\hspace{2cm}}}$$

3. Express the fraction as a probability. The chances that a person receiving Social Security benefits is *not* a retiree are about _____.

(**Note:** Since you already know that the chances are 7 out of 10 that a person *is* a retiree, you could also subtract $(10 - 7)$ to find that the chances are 3 out of 10 that a person is *not* a retiree.)

Does your statement sound something like this?

The chances that a person receiving Social Security benefits is not a retiree are **about 3 out of 10.**

How Can We Be Sure?

Can you correctly identify what type of beneficiary is approaching your desk simply by using probability?

No, you can't. Even if the chances are 9 out of 10 that a person is retired, you could never be sure that the person approaching your desk is not that 1 person out of 10 who *isn't* retired.

As you learned in the last two lessons, probability allows you to make *reasonable guesses*, not exact predictions.

For example, it is very hard to predict the outcome of one coin toss. (Will it come up heads or tails?) However, probability allows you to reasonably guess how many heads (or tails) would come up in 100 tosses.

Similarly, the data above cannot assure you that the person approaching your desk is a retiree. However, if 100 people receiving Social Security benefits approached your desk, you could make a reasonable guess that somewhere around 70 of them would be retirees.

Use the data to determine the probabilities below.

EXAMPLE   What were the chances that in 2010 a
registered voter in Kensington County
was an independent?

Registered Voters, Kensington County		
	2009	**2010**
Democrats	1,280	1,580
Republicans	960	820
Independents	320	600

STEP 1   How many 2010 registered voters were
there in all?

$1,580 + 820 + 600 = 3,000$

STEP 2   What fraction was registered independent?

$$\frac{600}{3,000} = \frac{1}{5}$$

ANSWER: The chances that a voter registered as an independent
in Kensington County in 2010 are **1 out of 5**.

4. What were the chances in 2009 that a voter registered in Kensington
County was a Democrat?

STEP 1   How many 2009 voters were there in all?

STEP 2   Write a fraction comparing the number of 2009
registered Democrats to all 2009 registered voters
in Kensington County. Reduce the fraction.

5. What was the probability that a voter registered in Kensington
County in 2009 was *not* a Republican?

6. What was the probability that a voter registered in Kensington
County in 2009 was an independent?

7. Was the probability that a voter registered in Kensington County
in 2010 was *not* a Democrat less than $\frac{1}{2}$ or greater than $\frac{1}{2}$?

8. Were the chances that a voter in Kensington County in 2010 was a
registered Republican less than $\frac{1}{3}$ or greater than $\frac{1}{3}$?

An Introduction to Sampling

Advertising Executive: "We took a survey of 1,500 people, and 1,125 of them said that they did not like our chocolate french-fry idea. So we've decided not to sell them at all."

French-Fry-Loving Friend: "Just because 1,125 people out of this entire country don't like the idea doesn't mean the rest of us don't! Why don't you listen to us instead?"

How do you think a company decides whether or not a product will sell well?

What mistake is the friend making in his comment above?

To the french-fry lover, 1,125 people does not seem like a large number—especially since he compares the number to the population of the whole country! However, the company is actually comparing **1,125** to the **1,500** people it surveyed—not all 300 million people in the United States.

Use what you have learned about percents and probability to complete the following statements.

The chances that a person would not like chocolate-covered french fries are 1,125 out of _____, or 3 out of _____.

The percent of Americans who do not like the idea of chocolate-covered french fries is _____ %.

In fact, based on the company's survey, the chance is **3 out of 4** that a person would not like chocolate-covered french fries—a total of **75%.** With so many people disliking the idea, the company was very wise not to go ahead and sell them.

But how can the company base its decision on the responses of 1,500 people? The company can use a **sample** of the population to predict how a larger group will respond.

Population and Sample

The study of statistics makes it possible to learn about a large group of people, places, or things (the **population**) by looking at information about a smaller part of that group (the **sample**). Look at some examples of how samples are used to make decisions and predictions every day in government, the business world, and even your own life.

EVERYDAY DATA

When choosing where to shop, people often compare prices in different stores. For example, suppose one store charges more for aspirin, paper products, toothpaste, and diapers. "This store has higher prices," you might say to yourself. "I'm not going to shop here." Based on a *sample* of prices, you made a decision about the prices in general, the *population*.

BUSINESS

A factory worker inspects a carton of light bulbs that has just been packed. She finds that 18 of the 240 bulbs are broken. "Something is wrong with the packing machine on this line," she says. "We'll need to recall all cartons that were packed today." Based on a *sample* of bulbs, the worker makes a prediction about what she'll find in all the cartons, the *population*.

GOVERNMENT AND POLITICS

Politicians often use public opinion statistics to get support for their policies and bills. Did the politicians ask each and every person in their district or state? Absolutely not. Instead, they chose a *sample* of the region, and based on how the voters in that sample answered, the politicians made accurate estimates of the opinion of the entire *population*.

For each situation given below, write the population and the sample.

EXAMPLE   To predict the total number of votes cast for each presidential candidate, a television station interviews voters and collects a sample of their preferences for president.

Population: _total votes_____   Sample: _voters interviewed___

1. An education official wants to know the standardized test scores for sixth-graders in his state. He looks at records from sixth-graders in six different towns in the state.

   Population: _____   Sample: _____

2. A candidate for mayor would like to find out how she is perceived among low-income residents in her city. She asks her staff to survey people living in city low-income housing units.

   Population: _____   Sample: _____

3. A shift supervisor counts defective aluminum canisters produced between 12:00 and 1:00 A.M. to discover how many defective canisters were produced on the entire early-morning shift.

   Population: _____   Sample: _____

4. A customer returns three poorly made toys to the same toy store and decides that none of the store's merchandise is of good quality.

   Population: _____   Sample: _____

5. A store that sells cookies offers free pieces of its chocolate chunk cookies to passersby. People who taste the cookies decide that all the baked goods in the store are delicious.

   Population: _____   Sample: _____

Random Samples

Optimistic Moviemaker: "I want to make a movie that will appeal to *everyone*—a real blockbuster. So, my partners and I interviewed more than 100 people, and the results were amazing. Over 95% of the people we interviewed wanted to see more horror movies—more violence, more monsters, more creepy stuff. It's obvious what kind of movie we should make!"

Money-Maker: "You didn't by any chance interview the 100 people leaving the 4 P.M. showing of that new horror picture downtown, did you?"

Optimistic Moviemaker: "Why, yes, we did. How did you know?"

What is the *population* being discussed above?

What is the *sample*?

Do you think most Americans want to see more horror movies?

To make accurate decisions and predictions based on statistics, you must use a sample that is **representative** of the population you want to learn about. Is this true of the sample taken above?

Put a check mark next to the categories of people *you* think would be likely to watch a weekday afternoon horror movie downtown.

Teenagers	City dwellers
Adults	Suburbanites
People who work during the day	People who like horror movies
People who do not work during the day	People who don't like horror movies

Did you find that the people the optimistic moviemaker interviewed were probably young people who were not working during the day and who generally like horror movies? The sample might include exceptions, but these would probably be the characteristics of most of the people interviewed.

What does this tell you about the sample chosen for the moviegoing population? Certainly, it was not representative of the public in general.

A representative group of Americans should include *at least*

- both males and females

- people of different ages

- people of different races and ethnic groups

- employed and unemployed people

- people from cities, suburbs, and rural areas

- poor, middle-class, and wealthy people

How Can We Get a Representative Sample?

There are different ways to obtain a representative sample of a population. The best-known and generally most accurate method is called **random sampling.**

Suppose you had a jar full of colored balls—some red, some blue, and some green—all exactly the same size. If you mixed them all up in the jar, closed your eyes, and pulled out a number of them, you would have a random sample of the balls in the jar. None of the balls is more likely to be pulled out than another.

A sample of a population is considered a random sample if each member of the population is equally likely to be chosen.

When a government, business, or organization wants information about a population, it uses random sampling to gather data. A population can be any group—not only adult humans. It can include

- all mammals living in Southeast Asia

- children under age five with the same flu virus

- graham crackers produced daily in New York State

Remember that *population* refers to *the group that you want information about.* To get this information, you look at a *random sample* within this population.

First find the population and the sample in each of the following situations. Then decide why each sample is *not* representative of the desired population.

1. A government official wants to find out what Americans think about cuts in the U.S. defense budget. He interviews 200 people outside a shopping mall in a small town near an army base.

   Population: _____ Sample: _____

   What is wrong with the sample? _____

   _____

2. A consumer wants to shop at the market with the freshest food. She reads the "date packed" label on whole chickens in five different stores.

   Population: _____ Sample: _____

   What is wrong with the sample? _____

   _____

3. "The people of the United States want a president who does not put a tax burden on people who have worked hard to achieve the American dream," says a candidate for office. The candidate's staff got this information from a survey of members of the American Small Business Owners Association.

   Population: _____ Sample: _____

   What is wrong with the sample? _____

   _____

4. A production worker inspects the first 100 circuit boards that are manufactured one Tuesday morning. She wants to know how many defective boards the manufacturer produces each month.

   Population: _____ Sample: _____

   What is wrong with the sample? _____

   _____

An Introduction to Surveys

- A candidate for public office wants to know whether or not the people in his state favor the death penalty.

- The personnel director at a large corporation wants to know if employees are happy in their jobs.

- A market research specialist wants to find out whether a company could make money manufacturing disposable socks.

What can these people do to get the information they need?

Government agencies, businesses, schools, political candidates, and nonprofit organizations are just some of the individuals and groups that conduct **surveys** or **polls** to obtain information. A *survey* or *poll* is a systematic collection of data from a random sample of a population.

Have you ever been involved in or heard about any surveys? What were they about? Have you ever phoned in your response to a televised survey question or completed an on-line survey?

_____

_____

_____

Surveys are usually conducted in one of three ways. Take a look at how each works.

Put a check mark next to the survey you would be most likely to fill out and return for each survey category.

1. **Mail or Internet**

    Surveys by mail or internet require you to fill out a questionnaire and mail/email it back to the sender.

    - Magazine music survey (favorite group, favorite song, and so forth)
    - Warranty and questionnaire that come with a product (stereo, appliance, and so forth)
    - Survey that comes with a cereal sample packet

    Would a self-addressed, stamped envelope or card influence your decision to respond to a survey? How? How about an on-line discount coupon or chance to win a sweepstakes drawing?

2. Telephone

Currently, there are two types of telephone surveys. One type requires people to make a phone call to record their opinion (usually an 800 or 900 number).

The other type features a polling group or market researcher calling random phone numbers and asking questions.

- A polling group asking whom you are going to vote for in the next election
- A 900 number to call to tell whether or not you support legalized abortion
- A market researcher calling to find out which household products you use

3. Personal Interview

Surveys are often conducted in person—in a shopping mall, on a street corner, or door to door.

- A survey conducted in a mall asking questions about you and your shopping habits
- A survey conducted on a street corner about what factors influence your choice of political candidates
- An official door-to-door census

Out of all the preceding choices, which survey would you be most likely to participate in? Why?

What do you think are the advantages and disadvantages of each type of survey?

1. Mail/Internet

Advantages: _____

_____

Disadvantages: _____

_____

2. Telephone

Advantages: _____

_____

Disadvantages: _____

_____

3. Personal Interview

Advantages: _____

_____

Disadvantages: _____

_____

Survey Organizations

You may already be familiar with the large survey and polling organizations in the United States. Here is a brief outline of some of them and what type of surveys they do.

PUBLIC OPINION POLLS

Major public opinion polling organizations include the *Gallup* poll, the *Roper* poll, and the *Harris* poll. These organizations survey people about their opinions in a wide range of areas, including politics, consumer products, and social issues.

They are highly respected organizations because they are *independent*. They do research and surveys on their own and sell their work to the media and other organizations that might be interested in reporting on it. These polling organizations provide objective data and statistics.

CURRENT POPULATION SURVEY

Like the U.S. Census, the *Current Population Survey* (or CPS) is a survey taken by the government, but it is taken *monthly*, not every ten years as is the census.

CPS is a survey of a random sample of 100,000 Americans, and it provides well-publicized information such as the unemployment rate, income statistics, and the birthrate.

NEWSPAPER AND NETWORK POLLS

Most major newspapers and television networks in the United States conduct surveys of their own.

Like Gallup, Roper, and Harris, the media also provide information about public opinion that they obtain from their own surveys.

4. Pick up a copy of a newspaper. Look for headlines and graphs that give statistics. Write down or cut out the headline, the date, the publication, the statistics, and *what polling or survey organization provided the statistics.*

Keep a journal of surveys and statistics over a period of time. This will provide an interesting year in review. It will also reinforce how frequently statistics are used.

Margin of Error

What do you think *margin of error* means in the chart at the right?

Percent of Olson City Citizens with Investments or Bank Accounts		
Type	African Americans	Whites
Savings Accounts	65%	75%
Checking Accounts	70	80
Stocks/Mutual Funds	30	40
Savings Bonds	20	30

Margin of error: ±2%

Source: *Olson City Journal*

Many statistics include a statement of **margin of error.** This information tells the reader that the numbers may vary from the actual population. When information from a sample is applied to a whole population, some allowance must be made for possible differences between sample and population.

For example, when you consider the margin of error, the percent of white Olson City citizens who have checking accounts is determined this way.

percent reported $\longrightarrow$ 80% − 2% = 78%
$\longrightarrow$ 80% + 2% = 82%

The percent of white Olson City citizens who have checking accounts is between 78% and 82%.

Use the data above to answer these questions.

1. What percent of African Americans in Olson City have savings bonds? _____ %

2. Considering the margin of error, what is the range of African Americans owning savings bonds?

    20% − 2% = _____ %

    20% + 2% = _____ %

    The range is from _____ % to _____ %.

More About Margin of Error

3. What can you say about the city council approval rating from March to April of 2010?

_____

_____

_____

_____

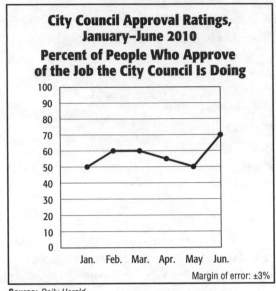

City Council Approval Ratings, January–June 2010

Percent of People Who Approve of the Job the City Council Is Doing

Margin of error: ±3%

Source: *Daily Herald*

4. What is the approval rating in March? _____ %

5. What is the March approval rating adjusted for margin of error?

from _____ % to _____ %

6. What is the approval rating in April? _____ %

7. What is the April approval rating adjusted for margin of error?

from _____ % to _____ %

8. Compare the low end of the range for March and the high end for April.

March: _____ % April: _____ %

9. Is it possible that approval *rose* between the two months?   Yes   No

At first glance, you may decide that the approval rating went down between March and April. However, *it is possible that the approval rating for the city council actually rose a slight amount between these two months.*

Acceptable margins of error are 5% or less. Otherwise, the statistics are considered inaccurate. Margins of error can also be written in decimal form. For example, ±2% would be shown as ±0.02.

Margins of error are a good reminder that the reporting of statistics is not an exact science.

In a random sample from West Valley, 1,500 teenagers were asked the following questions. Here's how they responded.

	Yes	No	Don't Know
"Do you have a close relationship with your parents?"	22%	71%	7%
"Are you happier now than you were five years ago?"	39	46	15
"Do you think that most of your friends are happy?"	35	55	10

Margin of error ±4%

Use the data above to answer the following questions.

10. What is the range of West Valley teenagers who do not have a close relationship with their parents?

from _____ % to _____ %

11. Considering the margin of error, is it possible that more teenagers surveyed *are* happier now than five years ago? What are the ranges of possible percents?

12. How many teenagers surveyed said that most of their friends are happy?

13. How many teenagers surveyed answered "Don't know" to each question?

EVALUATING DATA

Do Statistics Lie?

Used Car Dealer: "Numbers don't lie, I'm showing you the selling prices of all eight cars sold off this lot in the past week. The car you're looking at is priced below all of them! This car is a great deal."

Customer: "How do I know these figures are true? You could be making them up."

Used Car Dealer: "Here's the printout from our computer. Check the selling prices for yourself."

Customer: "If you're telling the truth, this car is cheaper than anything you've sold lately. I'll take it."

> Suppose the car dealer's numbers are accurate. What do they tell the customer about the car that's on the lot now?

> What conclusions is the customer jumping to using the data from the car dealer?

The car dealer used a list of numbers, or data, to try to sell a used car. But should the data really convince the customer? What does the purchase price of other cars tell you about the quality and price of this *particular* car?

The data that the dealer uses might be accurate, but the **conclusion** the customer draws from the data is *not valid*.

> Accurate data and statistics don't lie, but be careful of how the numbers are interpreted.

In this chapter, you'll get a chance to see how accurate and reliable data can be misunderstood and misinterpreted.

The practice you now have in making statements about data will help you detect conclusions that are not valid.

Here's the data shown to the customer on page 106. Suppose the customer is deciding whether or not to buy a 2005 Toyota Camry for $8,040. Make two statements about the data. One statement should compare the price of the Camry to the prices of these other cars.

1. The Camry _____

_____

_____

_____

Car	Purchase Price
'07 Chevy truck	$10,740
'08 Dodge truck	8,385
'05 Mazda RX7	16,900
'08 Nissan sedan	13,580
'06 Ford Focus	8,075
'07 Honda Accord	11,685
'06 Honda Civic	9,250
'05 Jeep Wrangler	10,655

2. _____

_____

_____

_____

_____

Your statements may sound similar to the ones below. Based on the data, were you able to say whether the Camry is a good buy?

- The Camry's price of $8,040 is lower than any other car price shown by the dealer.

- The purchase prices of the other cars range from $8,075 to $16,900.

The data cannot help you decide whether or not the Camry is a good buy. All you can really conclude is that the Camry is less expensive than any of the cars sold in the past week. As a consumer, you know that there is a lot more to buying a car (especially a used car) than comparing the price to the dealer's other prices!

This example shows that it is important not only to pay attention to statistics, but also to think about *how you interpret the statistics.*

3. **Critical Thinking** Was the car dealer lying? No. He provided accurate and truthful data. But was he being honest? Explain your reasoning.

Drawing Conclusions from Data

U.S. Leads World in Prison Use, a Survey Finds

LONDON—The United States has less than 5 percent of the world's population. But it has almost a quarter of the world's prisoners.

"These statistics prove what I've been saying all along," said Paul. "Our legal system in the United States is the best in the world. Just look at how many criminals our system arrests, convicts, and puts behind bars, compared to these other countries."

Number of People in Prison, per 100,000 Population—2008	
Country	Prisoners per 100,000
United States	751
Russia	627
England	151
WORLD MEDIAN	125
Germany	88
Japan	63

Source: International Center for Prison Studies, King's College, London, NYT, April 8, 2008

What do you think about Paul's statements?

Did he interpret the data accurately?

Do these statistics prove that the U.S. legal system is the best in the world?

Write a statement that summarizes the purpose of the data above. Remember to include information about *what, when, where,* and *how*.

Statement: _____

_____

_____

Does your statement sound something like this?

The chart shows the number of people in prison in 2008, per 100,000 population, in five different countries, as well as the median for the world.

At first, you may think Paul's statement makes sense. Maybe more prisoners do indicate a better system. But don't rely on just the numbers when you analyze Paul's statement! You need to use your thinking skills as well.

The data may be accurate, but the *conclusion* Paul draws from the data may not be accurate.

Asking yourself the following questions may help you draw other conclusions.

- What does it mean to have "the best legal system?" Does it mean that, when a crime is committed, the criminal is arrested, tried, convicted, and punished as swiftly and as inexpensively as possible?

- What information does this data give you about how many crimes are committed in the different countries and how quickly and inexpensively the criminals are imprisoned?

In fact, Paul's statement is not *proved* by the data. His statement is not necessarily untrue, but Paul needs other data to support his claim. A higher rate of people in prison is not necessarily "better" than a lower rate.

It is often difficult to decide whether or not someone's interpretation or conclusion is accurate. Always question how data is interpreted and what conclusions are being drawn.

Do not take it for granted that, if the numbers are accurate, any statement made *about* the numbers is accurate. *Your critical thinking skills are as important as the numbers.* The experience you now have in making statements about data will help you detect *invalid* conclusions.

The following two pages will allow you to practice using your thinking skills when analyzing data.

Each set of data below is accompanied by a statement that is not supported by the data.

a. Write a short explanation telling why each interpretation is *not necessarily an accurate conclusion.*

b. Then write a statement that is supported by the graph.

EXAMPLE "Wow!" says Jeanette. "People in Watertown must get really dirty in August. Just look at how much soap they buy!"

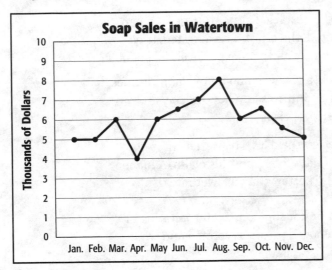

a. The fact that soap sales are high in August does not necessarily mean that people actually need more soap then or that they are dirtier.

b. Soap sales in Watertown were highest in August and lowest in April.

1. "Numbers don't lie. Oral contraceptives are the best method for birth control if you don't want to be sterilized."

a. _____

_____

_____

b. _____

_____

_____

Failure Rates of Most Common Contraceptive Methods
Sterilization—less than 1%
Oral Contraceptives—6%
Condoms—16%
Diaphragms—18%

Source: Alan Guttmacher Institute

2. "As the graph indicates, the citizens of North County pay far too much for police protection. Our police officers' salaries are the highest in five counties!"

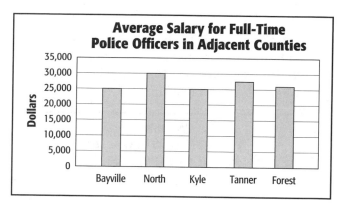

Average Salary for Full-Time Police Officers in Adjacent Counties

Dollars: 0, 5,000, 10,000, 15,000, 20,000, 25,000, 30,000, 35,000

Bayville   North   Kyle   Tanner   Forest

a. _____

_____

_____

b. _____

_____

_____

3. "Look at these statistics, Bob. It just goes to show you—our fire protection is not as great as you say it is."

a. _____

_____

_____

_____

Oldbrook Fire Deaths, 2010	
Some Causes of Fire Deaths	Percent
Smoking	29
Electrical	27
Candle/Incense	9
Arson	7
Cooking	5

b. _____

_____

_____

Misinterpreting Graphs

Suppose you spoke to two job counselors.

First Counselor (Graph A): "As you can see from the graph, wages for production workers have been rising at a terrific rate. You'd be doing yourself a favor by getting into this field."

Second Counselor (Graph B): "The figures indicate that production wages are pretty flat. You might want to look into a field where wages are rising more quickly."

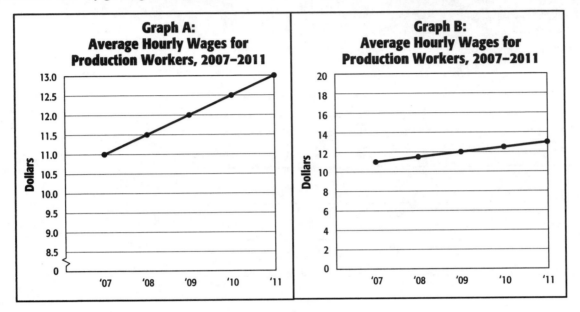

Which person is telling you the truth? Are they both reading the data correctly?

How can you make a decision based on two graphs that seem to be contradicting each other?

Fortunately, your work in this book and your own critical thinking skills will help you interpret graphs that might be misunderstood.

Make a statement summarizing Graph A. _____

_____

Does your statement hold true for Graph B?   Yes   No

Although Graph A seems to show a steady rise in wages and Graph B seems to show little rise, **both graphs actually present the same data.**

To see this clearly, fill in the blanks below. Use one graph to answer the question and the other graph to check your answer.

Between 2007 and 2011, production workers wages rose from about _____ per hour to about _____ per hour.

Both graphs show a rise from about **$11.00 to $13.00 per hour.**

If both graphs use the same data, why do the graphs look so different? The difference is in **scale.** Scale is the increase in value along the axis of a graph. As you learned in Organizing Data, the scale used to present data can vary. Sometimes the values on the scale can be multiples of 2, sometimes multiples of 5, 10, or 100.

The scale on Graph A increases by _____ at each interval.

The scale on Graph B increases by _____ at each interval.

With a smaller scale (increases of $0.50), the data line on Graph A rises higher to reach the $2.00 wage increase. With a larger scale (increases of $2.00), the data line does not need to rise very much at all.

In addition, compare the spacing between the dollar amounts on the scale of Graph A to the spacing of the dollar amounts on Graph B.

Which graph has larger spacing between the values on the vertical axis?

The spacing between the values on Graph A is greater than the spacing on Graph B. Therefore, the data line is steeper on Graph A.

So, how can you tell, by looking at the graphs, whether wages have increased a lot or a little?

You decide. You learned in Understanding Data that a number is neither small nor large until you compare it to another number. Neither Graph A nor Graph B will tell you whether $2.00 is a large increase. Understanding the data, collecting more information, and *using your thinking skills* will help you decide.

Which of the following will help you decide whether a wage increase of $2.00 per hour is large or small? Circle Yes or No for each.

- the hourly wage increase in a related industry   Yes   No
- the number of jobs in the production industry   Yes   No
- the salary of a manager in the production industry   Yes   No

Did you decide that it would be useful to know how wages increased in related industries? If so, your thinking skills are right on target. Although you might be interested to know how many jobs are available and what a manager's salary is, these figures will not help you evaluate the $2.00 wage increase.

Your work in this lesson should help you keep in mind two important things.

1. *When analyzing data on a graph, pay close attention to the scale.* Don't look at just the shape of the data line. Use the skills you've developed in making statements about data to determine what is happening on the graph.

2. *Use your thinking skills.* Don't rely on what someone else says about the data. Find appropriate numbers to compare with the data. Use your judgment in deciding about the size of a figure.

Use the graphs and statements to answer the questions.

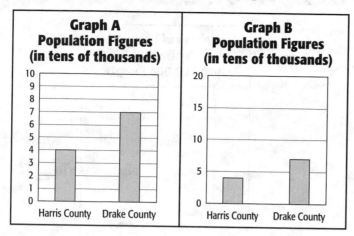

Statement 1: "The graph shows how much greater the population is in Drake County than in Harris County."

Statement 2: "As you can see, there is not a large difference between Harris County and Drake County in terms of population."

1. Which statement do you think was made about Graph A? Which one was made about Graph B?

   a. Statement 1 refers to Graph _____.

   b. Statement 2 refers to Graph _____.

2. Make a statement comparing the population of Harris County with the population of Drake County.

   _____

   _____

3. Decide whether or not each of the following numbers would be useful in discovering if there is a big difference in population between the two counties. Circle Yes or No.

   Yes   No   **a.** the population figures for other counties in the state

   Yes   No   **b.** the number of people per square mile in each county

4. Write a paragraph describing the scale in each graph and how the scale influences the height of the data bars.

The Importance of Zero on a Scale

"Wow!" said Myra. "Look at this. The burglary rate in the South is about twice the rate in New England. I didn't realize there was such a big difference."

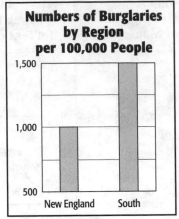

Numbers of Burglaries by Region per 100,000 People

Source: Uniform Crime Reports, FBI

Myra is correct that the bar showing the South's burglary rate *looks* about twice the height of the New England bar. But let's look at the data more carefully. Fill in the blanks below.

- There were _____ burglaries per 100,000 people in New England.

- There were _____ burglaries per 100,000 people in the South.

Is 1,500 twice as large as 1,000? Of course not. Then why does the graph make it appear that way?

The graph above may be misleading because it does not have a *zero* on the vertical scale. To see how important a zero is to the scale, compare the graph above to the one below—a graph that *does* include the zero.

As you can see, by including the full length of each bar (including the values between 0 and 500 that were left off the graph above), the graph pictures the correct proportion between 1,000 and 1,500.

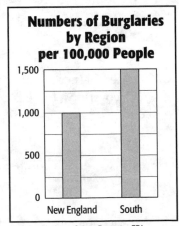

Numbers of Burglaries by Region per 100,000 People

Source: Uniform Crime Reports, FBI

It is important to look not just at the *picture* of the graph but also at the *labels* and *scale*. Beware of graphs that do not show a zero on the scale. Chances are that the proportions on the graph are misleading.

Zero Break

Sometimes you will see a graph that does show a zero on the scale, but then shows a *break* in the axis. The graph below is an example. A break like this is called a **zero break** or a **notch.**

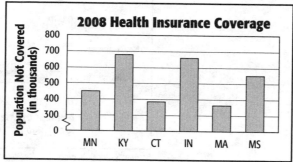

Source: U.S. Department of Commerce

Because all the data points to be included on the graph were above 300,000, the graph focuses on those values.

Is this graph misleading? Whenever you see a graph that includes a zero break, remember that the scale of the graph is different than it would be if all the values were included.

A graph showing a zero break is better than a graph that shows no zero on the scale at all.

To construct a graph using a zero break

STEP 1 Find the lowest and the highest values you will need to graph; mark these on the vertical axis.

For example, on the graph above, the lowest value to be graphed was about $350,000, and the highest was about $650,000.

STEP 2 Below the lowest value on the vertical axis, put a slash to indicate a break in the axis. Under the slash, write in a zero.

STEP 3 Finish graphing as usual.

Use the data for each problem to answer the questions.

1. "The number of live births in New Hampshire is half the number of live births in West Virginia according to this graph."

   Is this an accurate reading of the data? Why or why not? _____

   _____

   _____

   _____

   _____

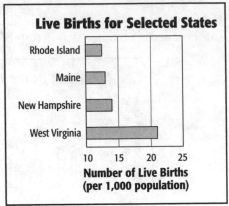

Live Births for Selected States

Rhode Island
Maine
New Hampshire
West Virginia

10   15   20   25

Number of Live Births (per 1,000 population)

Source: National Center for Health Statistics

2. Redraw the graph so that it is not misleading. Remember to include a zero on the scale.

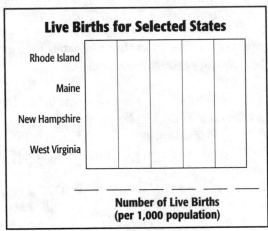

Live Births for Selected States

Rhode Island
Maine
New Hampshire
West Virginia

— —   — —

Number of Live Births (per 1,000 population)

Source: National Center for Health Statistics

3. Use the data below to construct a line graph. Use a zero break between 0 and $1.25.

Average Cost of a Fast-Food Hamburger
2005: $1.25
2006: $1.35
2007: $1.50
2008: $1.75
2009: $1.70
2010: $1.60

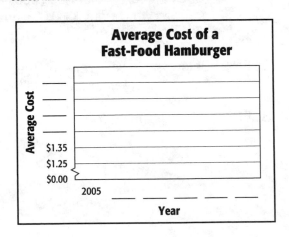

Average Cost of a Fast-Food Hamburger

Average Cost

$1.35
$1.25
$0.00

2005

Year

Understanding Correlation

Think back to what you learned about scatter diagrams in Organizing Data. Is there a correlation between hours spent watching TV and hours of free time?

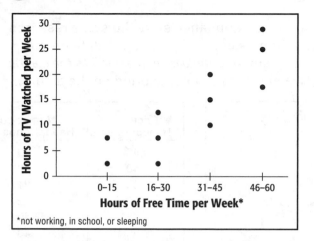

The scatter diagram above shows that there *is* a **correlation** between the two values graphed. As the number of hours of free time increases, so does the number of hours spent watching TV.

EXAMPLE 1 People with 25 hours of free time per week watch _____
(fewer, more)

hours of television than people with 10 hours of free time.

You should have written **more** in the space above.

The relationship shown above is a **positive correlation** because as one value rises, the other rises. To show correlation on a scatter diagram, a diagonal line is often drawn to show how most values cluster in a general pattern.

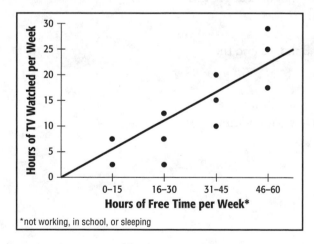

Another kind of correlation is a **negative correlation**. In this relationship, as one value rises, the other falls. Let's look at an example.

EXAMPLE 2   A manager in an appliance store did some research into headphone sales and plotted the data on this scatter diagram. *Add the data in the chart to the diagram.* Other data has already been plotted on the diagram.

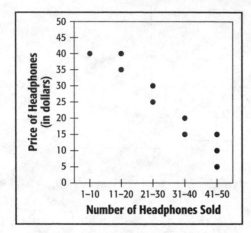

Price of Headphones	Number of Headphones Sold
$45	10
40	10
35	21
30	22
25	40
20	40

What is the relationship between the two values on the diagram?

As the price of the headphones ———————————————, the
                                              (rose, fell)

number of headphones sold ———————————————.
                                    (rose, fell)

Did you write **fell** in the first blank and **rose** in the second blank? It is also true that as the price of the headphones rose, the number of headphones sold fell.

There is a correlation between the price of stereo headphones and the number sold; it is a negative correlation.

As you learned in Organizing Data, some values may not follow the pattern. Remember that a correlation can exist even if a few data points do not fall within the pattern.

Correlation vs. Cause

A recent consumer survey produced the data at the right.

"Hey, look at this," said Kathleen. "The more often people eat out, the more often they go to see a doctor. Information like this makes you never want to eat in another restaurant!"

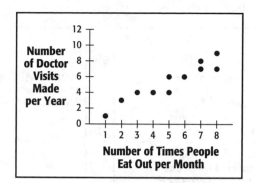

Is there a correlation between the number of times people eat out and the number of doctors' visits they make?

Which of the following do you think is the cause?

- Frequently eating in a restaurant *causes* people to see a doctor more often.

- More frequent doctor visits *cause* people to eat out more.

- Neither of the above.

Actually, although there is a correlation between these two values, neither one is the **cause** of the other. Kathleen jumped to a faulty conclusion when she analyzed the data. She did not understand that **correlation** is not the same thing as **cause**.

If eating in restaurants does not cause more frequent visits to the doctor, why does this correlation exist?

In fact, trips to the doctor and trips to a restaurant increase as a third value increases—income. The more money people make, the more they eat out in restaurants and go to the doctor. People with less money tend not to seek frequent medical care or eat out often because of the expense.

Sometimes, though, when there is a correlation between two values, one *can* be the cause of the other. In other words, one value can **have an effect on** the other value.

---

1. **Critical Thinking** There is a negative correlation between a person's chances of being murdered and his or her income. In other words, the lower a person's income, the more likely he or she will be murdered. Why do you think this is so?

Circle the value that has an effect on the other value in each of the following correlations.

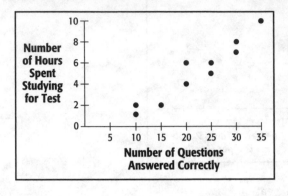

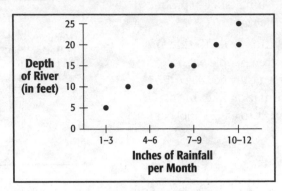

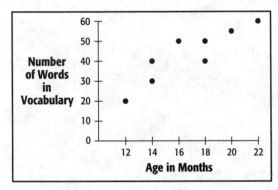

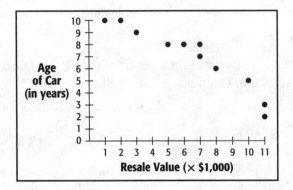

You should have found that

The number of hours spent studying had an effect on the number of right answers.

The amount of rainfall had an effect on the depth of the river.

Age had an effect on vocabulary.

The age of a car had an effect on its resale value.

These four examples show a causal relationship between values. Remember, however, that *not all correlations show causation* (cause).

Use the data below to answer the questions that follow.

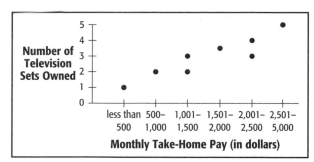

2. a. There is/is not a correlation between

_____ and _____.

   b. "Now that I've seen these figures, I'm going to go out and buy another television set! My income is sure to go up."

      Is the analysis accurate? Why or why not?

_____

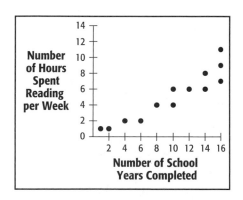

3. a. As the number of school years completed _____, the number of hours spent reading per week _____.

   b. There is a _____ correlation between school years completed and time spent reading.

Bias in Surveys and Statistics

Jean: "This newspaper article says that over 60% of Americans approve of using our armed forces in the growing conflict overseas. That's 3 out of every 5 people! With a majority like that, I can't blame Congress for voting the way it did."

Roberto: "That's funny. A flier someone handed me on my way to work said that almost 80% of Americans did *not* approve of getting involved in the war. What's going on?"

Have you ever come across conflicting statistics?

How can two sources like the ones above give such differing figures?

As you learned, some survey organizations are **independent** groups; in other words, they have no financial or political interest in their survey results. In general, their methods are respected and accurate.

Some surveys, however, are not as reliable. On page 96, you saw the inaccurate results of a survey conducted with a "nonrandom" sample—the horror movie-goers. Even though a group may intend to conduct a fair survey, flawed methods likely produce inaccurate statistics.

In addition, some polling groups have a **bias** concerning the outcome of a survey. In other words, they have a **preference** as to how people respond to questions on the survey.

If a survey's results appear to contradict other statistics, it is a good idea to look at the **source**—the group that gathered and analyzed the data. Ask yourself whether the source might benefit from people believing its survey results.

Suppose that the figures on the flier Roberto received were compiled by one of the groups below. Circle the group you think provided the statistics.

People for Peace, an organization that lobbies heavily against armed force involvement

Make America #1, a group that promotes the use of U.S. military power in conflicts worldwide

People for Peace would probably benefit if you believed that almost 80% of Americans were against getting involved in the war. These statistics support its cause. Make America #1 would probably have no interest in reporting that 80% of Americans did not support its point of view.

Of course, People for Peace may be reporting valid and accurate statistics. You simply cannot tell just by looking at them. However, if a source shows the possibility of bias, it is a good idea to find statistics from other sources to compare.

1. **Critical Thinking** Suppose the statistics on Roberto's flier came from an independent polling agency. Would you be more likely to believe them? Why?

Put a check mark next to each survey conducted that you think might show bias. Ask yourself if the people conducting the survey might have a preference for what the survey results showed.

2. A group called Citizens for Smith conducts a survey to find out how many people plan to vote for Smith in the city council election.

3. A soft-drink company reports a survey revealing which brand of soft drink teenagers prefer.

4. Americans Tough on Criminals surveys crime victims on their views concerning capital punishment.

5. A newspaper polls local communities to find out how they feel about gun control.

6. An inventor, trying to get financial support for her idea, does a survey asking people if they would buy a computerized laundry basket.

7. A television network conducts an exit poll outside a voting center to project the winners in an election.

Biased Methods

How might an organization go about gathering statistics that support its point of view? To be sure of getting the *right* responses, the polling group could use several different methods. Let's take a look at a couple of them.

SAMPLE BIAS

To obtain the results it wants, a group could survey only certain people—people who are likely to have the same point of view as the survey group. You saw an example of sample bias on page 96 in the survey of moviegoers. Sample bias is sometimes done intentionally and at other times by mistake.

As you do the activity below, remember that Make America #1 is the organization that promotes the use of military power.

Which of the groups below would Make America #1 most likely want to survey to find support for its cause?

a community with a large military base

a university town known for its antiwar demonstrations

You probably know that the first community would be the best choice for Make America #1. Although not *everyone* in the first community would be in favor of military involvement, there would certainly be a higher percent there than in the second group!

This example demonstrates the importance of a **random sample.** Almost any organization can find and survey a group that is known to support its positions. However, the statistics obtained would not necessarily reflect the opinion of most people.

Statistics are often accompanied by an explanation of the random sample and how it was obtained. If you suspect that the statistics are not based on a random sample, you should consider the possibility of bias and look for other sources and studies on the same topic before drawing conclusions.

QUESTION BIAS

Another way that survey groups can affect the outcome of their polls is by asking leading, or loaded, questions. This is called **question bias.** The wording of this type of question actually *directs* the respondent to answer in a certain way.

Go back and read the dialogue between Jean and Roberto on page 124. Both statistics have something to do with military involvement, but are they really contradicting each other?

Suppose the 60% quoted in Jean's article answered yes to the following question:

Do you approve of using our armed forces in the growing conflict overseas as a peacekeeping force to protect the innocent citizens of both countries?

And suppose that the 80% quoted in Roberto's flier answered no to this question:

Do you approve of getting involved in the war overseas by sending in armed American troops to fight in the battles of other countries?

There is a big difference in the numbers quoted because there is a big difference in the questions asked.

In addition, many survey questions are designed so that the person answering the question must choose from a multiple-choice format. But what happens if a person's true opinion is not listed as one of the choices? Many times, the answers recorded in a survey are not an accurate representation of people's true opinions.

---

8. **Critical Thinking** Circle Yes or No to the following questions.

Are taxes a good thing?	Yes	No
Is America a free country?	Yes	No
Are Americans hard-working people?	Yes	No

You may have found yourself wanting to answer "not always" or "yes, but . . ." Can you see that yes/no type questions often do not allow you to express your true opinion?

The two organizations below surveyed people concerning their opinions on the U.S. space program.

Each group asked a loaded question to get a response that matched its own point of view. Match the organization to the question you think it would ask.

Space Is Our Future supports increased spending for space exploration.

"Are you tired of our government spending millions of dollars on useless efforts in space exploration while our own air, water, and plant life on Earth is in jeopardy?"

Home Earth seeks to divert government funds from space exploration toward environmental programs here on Earth.

"Do you believe that our government should put more resources into vital technological advancements such as space exploration so as to maintain our scientific superiority in the world?"

Did you match Space is Our Future with the second question and Home Earth with the first question?

Can you see how it would be difficult to answer no to either of these questions? The wording of each question leads us to answer yes. The survey statistics obtained from the responses to these two questions would be very different, wouldn't they? By using biased questions, each group could claim support for its own position.

Decide if each of the following survey situations is biased or unbiased. Write a brief explanation telling why.

9. To find out how city residents felt about a new factory being built in the midst of a large neighborhood, the chairman of Citizens for Economic Development asked, "Do you agree that new jobs and more taxpaying industry are important developments in this area of economic frailty?"

   biased   or   unbiased

   Why? _____

   _____

   _____

10. A television network surveyed people to find out what percent planned to vote in an upcoming presidential election.

biased   or   unbiased

Why? _____

_____

_____

11. To find out how city residents felt about a new factory being built in the midst of a large neighborhood, a group called Save Our Homes asked, "Are you willing to sacrifice your homes and community to a factory that will destroy the environment?"

biased   or   unbiased

Why? _____

_____

_____

12. A researcher surveyed people outside an exclusive shopping mall to find out how Americans were affected by the current recession.

biased   or   unbiased

Why? _____

_____

_____

13. A consumer opinion researcher stopped people coming out of an ice cream parlor and asked them what their favorite dessert was.

biased   or   unbiased

Why? _____

_____

_____

Critical Thinking with Data

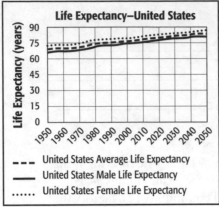

Life Expectancy—United States

Source: Data 360

Circle the statement(s) that are true based on the graph above.

- **Statement 1:** Average life expectancy in the United States is expected to rise until 2050 due to improved hospital care.

- **Statement 2:** There has been a huge increase in average life expectancy in the U.S. between 1950 and 2050.

- **Statement 3:** During the early 2000s, people in the United States were more healthy than in the mid 1950s.

- **Statement 4:** Average life expectancy for both males and females living in the United States is expected to rise between 1950 and 2050.

An analysis or interpretation often accompanies data. Don't *assume* the analysis is accurate—use your **critical thinking** skills to see if the analysis is supported by the data.

At first, each statement above might appear true based on the graph. Let's find out why only statement 4 is actually based on the data provided.

For each statement, think about these questions.

- Is anything missing?
- Can the data be interpreted differently?

Statement 1: Average life expectancy in the United States is expected to rise until 2050 due to improved hospital care.

Is anything missing? Yes, the graph gives no information about the causes of the increase. You do not know if improved hospital care was the cause.

Remember: Even if you believe the statement to be true, the question asks which statements are *supported by the graph,* not which ones are true based on your prior knowledge or opinion.

Statement 2: There has been a huge increase in average life expectancy in the U.S. between 1950 and 2050.

Can the data be interpreted differently? How? (**Hint:** What does *huge* mean?)

_____

_____

_____

The graph shows clearly that there was an increase in life expectancy. But how can you decide whether this growth was "huge?"

In Understanding Data, you learned to compare numbers. You learned that a number is neither small nor large *until you compare it to another number.* In the example, what is the increase being compared to? Life expectancy in the 1800s? Life expectancy in another part of the world?

Statement 3: During the early 2000s, people in the United States were more healthy than in the mid 1950s.

Is anything missing? Can the data be interpreted differently?

_____

_____

Longer lives do not necessarily mean healthier lives. People may in fact be living longer in an unhealthy state, kept alive by machines and/or drugs. To allow you to make a statement like this, the graph would need to show data concerning other aspects of health.

Statement 4: Average life expectancy for both males and females living in the United States is expected to rise between 1950 and 2050.

This statement is true based on only the data provided on the graph.

Let's look at another example. Circle the statement(s) that are true based on the graph.

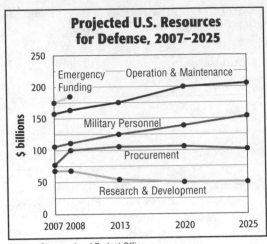

Projected U.S. Resources for Defense, 2007–2025

Source: Congressional Budget Office

- **Statement 1:** The United States will continue to spend more money on defense than on education and social welfare programs.

- **Statement 2:** Between 2013 and 2025, the U.S. is projected to increase spending on military personnel and operation and maintenance.

- **Statement 3:** More money will be needed for personnel than for research and development because the U.S. military is already highly paid.

- **Statement 4:** In 2025, 50 billion dollars for research and development will most likely not be adequate to preserve the strength of the U.S. military.

Statement 2 is the only statement that can be made based on the graph.

Let's look at the other statements more closely.

Statement 1 cannot be made based on the graph because there is no information given regarding education and social welfare spending. Though the reader may believe the statement to be true, one cannot conclude it based on this graph alone.

Statement 3 cannot be made based on the graph because the graph gives no *reasons* for how resources are spent.

Statement 4 cannot be made based on the graph. Though 50 billion dollars is in fact the amount to be spent on research, the graph does not tell us whether this amount is adequate.

Answer the questions using the graph below.

1. Make a statement about the number of Richmond homicides during 2010.

_____

_____

_____

_____

_____

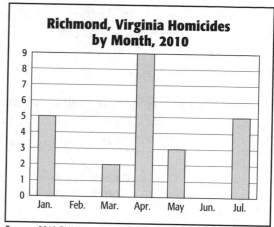

Richmond, Virginia Homicides by Month, 2010

Source: *2010 RVANews*

2. Which of the following statements can be made based on the graph?

   a. Between January and July 2010, there were 240 homicides in Richmond, VA.

   b. More children died by homicide in April 2010 than in January 2010.

   c. There were no deaths in Richmond, VA in February and June 2010.

   d. 2010 homicides in Richmond, VA ranged from a low of 0 to a high of 9 between January and July.

Posttest A

This posttest gives you a chance to check your skill at data analysis. Take your time and work each problem carefully. When you finish, check your answers in the back of the book and review any topics on which you need more work.

For problems 1–3, refer to the following data.

Davco, Incorporated			
Department	**Number of Employees Who Carpool to Work**	**Number of Employees Who Take Public Transportation**	**Total Number of Employees in Department**
Sales	26	55	110
Service	10	10	25
Administrative Pool	15	12	30
Maintenance	10	6	18
Management	1	5	20

1. Which of the departments, sales or service, has the higher *rate* of employees who carpool to work?

2. According to the chart, about 30% of Davco employees carpool to work. What numbers can you compare with this figure in order to decide if 30% is a large number or a small number?

   a. the total number of employees

   b. the percent of people who carpool to other nearby industries

   c. the number of cars in the company parking lot

3. Write a statement about the number of management employees who carpool to work compared to the number of maintenance workers who carpool.

For problems 4–7, refer to the following data.

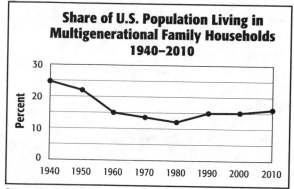

Source: Pew Research Center analysis of the U.S. Decennial Census data

4. Which of the following best summarizes the graph?

   a. The graph shows the percent of U.S. population between 1940 and 2010 who lived in a multigenerational household.

   b. The graph shows the decline in the number of people living in a multigenerational household between 1940 and 2010.

   c. The graph shows the percent of people in the United States who lived in a multigenerational household.

5. What is the range of percents provided on this graph?

6. Use the data to make a statement comparing the percent of the U.S. population living in multigenerational households for two different years.

7. Which statement about the graph is *not* true?

   a. The percent of the U.S. population living in multigenerational housing in 2010 was higher than in 1980.

   b. The percent of the U.S. population living in multigenerational households steadily declined between 1940 and 1980.

   c. In 1990, 20 percent of the U.S. population was living in multigenerational households.

For problems 8–11, refer to the following newspaper article.

According to a recent survey, over 70% of Americans would rather eat a baked potato than rice at mealtimes, and 20% actually eat potatoes for breakfast. At the annual Potato Lovers Picnic in Daisy County last year, almost 750 pounds of potatoes were consumed by the 3,000 participants there. A spokesman for the event reported that potato sales are up 5% in the county from last year's sales figure of $940,000.

8. Which of the following organizations would most likely benefit from biased statistics in the survey above?

   a. WXYZ television news

   b. American Food Industry, Inc.

   c. American Potato Growers, Inc.

9. To ensure that people say that they prefer potatoes to rice, which of the following questions might the survey group ask?

   a. Would you prefer a hot, buttered baked potato or a bowl of rice?

   b. Which do you prefer to eat: a baked potato or rice?

   c. Which do you have more often at mealtimes: a baked potato or rice?

10. Which of the following would be a biased sample on which to conduct the potato-or-rice survey?

   a. 1,000 people working in a city government building

   b. 500 people leaving a professional baseball game

   c. 3,000 participants at the Potato Lovers Picnic

11. According to the survey, what are the chances that an American will eat potatoes for breakfast?

For problems 12–14, refer to the following data.

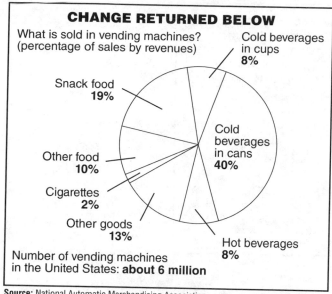

CHANGE RETURNED BELOW

What is sold in vending machines?
(percentage of sales by revenues)

Cold beverages
in cups
8%

Snack food
19%

Cold
beverages
in cans
40%

Other food
10%

Cigarettes
2%

Other goods
13%

Hot beverages
8%

Number of vending machines
in the United States: **about 6 million**

Source: National Automatic Merchandising Association

12. Which category of vending machine sales had the greatest percent?

13. What is the purpose of the graph above?

   a. to show how much change people receive from vending machines

   b. to show the rising cost of snack foods sold in vending machines

   c. to show percentage of vending machine sales for different categories

14. Use the graph to write a statement about vending machine sales.

For problems 15–18, refer to the following data.

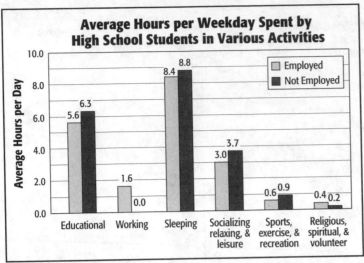

Source: Bureau of Labor Statistics

15. How many hours per weekday do employed high school students spend working and sleeping?

16. Which of the following statements summarizes the data?

   a. The graph shows the number of hours on average high school students spend engaged in different activities on weekdays.

   b. The graph shows the percent of both employed and unemployed high school students who engage in different activities.

   c. The graph shows the number of high school students who engage in different activities when working and not working.

17. Use the graph to write a statement comparing two different activities.

18. Which of the following statements is *not* true according to the graph?

   a. Unemployed high school students spend about 3.7 hours per weekday socializing and relaxing.

   b. On weekdays, employed high school students spend less time on educational activities than unemployed high school students.

   c. Unemployed high school students spend 1.5 hours more on exercise and sports than employed high school students.

For problems 19–22, refer to the following data.

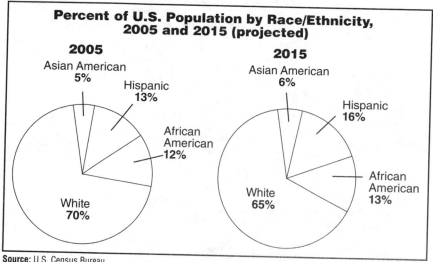

Percent of U.S. Population by Race/Ethnicity, 2005 and 2015 (projected)

2005
- Asian American 5%
- Hispanic 13%
- African American 12%
- White 70%

2015
- Asian American 6%
- Hispanic 16%
- African American 13%
- White 65%

Source: U.S. Census Bureau

19. What percent of the 2005 U.S. population was Hispanic?

20. Which of the following statements best summarizes the data of both graphs?

   a. There are more whites living in the U.S. now than ever before.

   b. The percentage of whites in the U.S. is projected to decrease between 2005 and 2015, while the percentages of other racial groups will increase.

   c. As the percentage of whites in the U.S. decreases to 65%, so too will the overall U.S. population.

21. Write a statement about the population of the U.S. in the future.

22. Use data from both graphs to *estimate* the percent of the U.S. population that will be Hispanic in the year 2012.

For problems 23–25, refer to the following data.

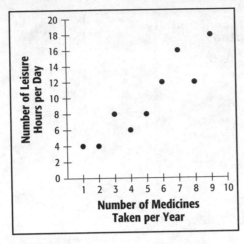

23. Is there a correlation between number of leisure hours and number of medicines taken?

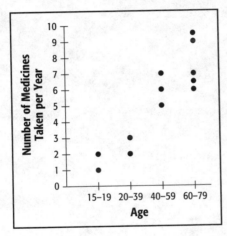

24. Is there a positive or a negative correlation between age and number of medicines taken?

25. Write a short summary of the data above. Do you think any of the three factors listed (age, leisure time, number of medicines taken) has an effect on the other factors?

For problems 26–28, refer to the following data.

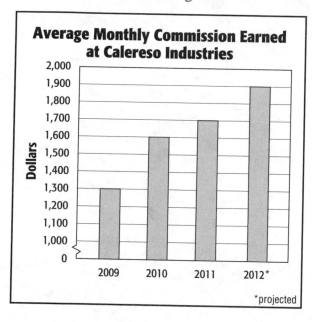

Average Monthly Commission Earned at Calereso Industries

26. Write a statement about average monthly commission at Calereso Industries in 2012.

27. Which of the following statements is true according to the graph?

   a. The average monthly commission at Calereso will exceed $2,000 in 2013.

   b. The average monthly commission at Calereso was twice as high in 2010 as in 2009.

   c. The average monthly commission at Calereso rose steadily between 2009 and 2011.

28. Was the average monthly commission earned in 2011 twice the amount earned in 2009? Why does it appear that way on the graph?

For problems 29–32, refer to the following data.

Projected Election Results	
Crane	46%
Davidson	39%
Thurman	10%
Rizzo	5%

Margin of error: ±4%

29. Given the margin of error, what is the range of percent of projected votes received by Thurman?

30. Considering the margin of error, is it possible that Crane will end up with half of the votes?

31. If 400,000 votes were cast for these four candidates, *approximately* how many were cast for Crane?

32. Given the margin of error in this poll, is it possible that Davidson will win the election?

For problems 33–35, refer to the following data.

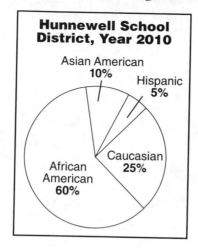

Hunnewell School District, Year 2010

Asian American 10%

Hispanic 5%

Caucasian 25%

African American 60%

33. What were the chances that a child in the Hunnewell School District in 2010 is African American?

34. Write a statement comparing two racial groups in the Hunnewell School District in 2010.

35. If there were 12,400 students in the Hunnewell District in 2010, how many were Hispanic?

POSTTEST A REVIEW CHART

Circle the number of any problem that you miss. A passing score is 30 correct answers. If you passed the test, go on to Using Number Power. If you did not pass the test, review the chapters in this book.

Problem Numbers	Skill Area	Practice Pages
1, 2, 3, 6, 14, 17, 21, 34	understanding data	8–25
4, 5, 7, 12, 13, 15, 18, 35	organizing data	26–57
8, 9, 23, 24, 25, 28	evaluating data	106–133
10, 11, 29, 30, 32, 33	using probability and statistics	86–105
16, 19, 20, 22, 26, 27, 31	using data sources	74–81

USING
NUMBER
POWER

A Look at the U.S. Census

Every 10 years, the federal government collects, tabulates, and publishes statistics about the American people. These statistics make up the **U.S. Census.** The census is used, among other things, to determine

- how many representatives each state should have in Congress
- how much money each city and state should receive in federal aid
- how much money to allocate for bilingual education
- the need for low-income housing

The census attempts to count *each and every person* living in the United States. Just imagine what a data collection task that is—considering that there are over 310 million people living in the United States today! To get information from all these people, the government sends out questionnaires to home addresses. Those people who do not respond to the questionnaires are reached by door-to-door census takers. An estimated 565,000 civil servants were needed to collect and process the data for the 2010 census. Even more will be needed for the year 2020 census.

Because the census is such an enormous project, some people slip through the cracks and are not counted. This results in an *undercount*—a number below the actual total. Due to some serious undercounts in the 1990 census, efforts have been made to improve the accuracy the census in following years.

What did the 2000 census tell us about the people of America? Here are some statistics.

Age
The nation's median age is 35.3 years. This is up from 32.9 years in 1990.

Gender
Of the population, 50.9% is female. This number is down from 51.3 in 1990. Interestingly, however, at age 85, there were twice as many women as men!

Households

Of the U.S. households, 52% are maintained by married couples. This is down from the 1990 figure of 55%.

Persons living alone account for 26% of households, up from 25% in 1990.

Where We Live

Of the U.S. population, 80.3% live in metropolitan areas.

The Los Angeles/Anaheim/Riverside metropolitan area experienced a 12.7% population growth, while the New York/New Jersey/Long Island/Connecticut area experienced an 8.4% growth.

The chart below contains some more data from the U.S. Census. Use the data to answer the following questions.

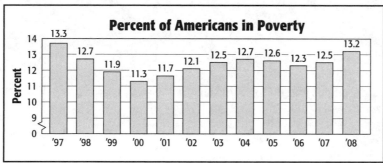

Percent of Americans in Poverty

Source: U.S. Census Bureau

1. In what year was the percentage of Americans in poverty the lowest between 1997 and 2008?

2. What is the range between the highest and lowest percent of Americans living in poverty between 1997 and 2008?

3. **Critical Thinking** Census data can be collected by mail, phone, and door-to-door interviews. Which person would be less likely to be counted: a homeless person or a homeowner? What effect do you think this has on the census statistics?

Are You Average?

As you know, comparing data is a good way to learn more about a subject. In this lesson, you will compare the amount of money you spend on certain items to the **average** expenditures among different groups in the United States. By doing this, you can learn *where you fall* among national averages.

First, estimate how much money you spend *per year* in the following categories.

To do this, multiply your *average* monthly expenditures by 12 or your *average* weekly expenditures by 52.

For example, if you spend about $80 per week on food,

$$\begin{array}{r} \$80 \\ \times\ 52 \\ \hline \$4{,}160 \end{array}$$

your *annual food expenditure* would be $4,160.

And if you spend about $230 per month on gas and oil for your car,

$$\begin{array}{r} \$230 \\ \times\ 12 \\ \hline \$2{,}760 \end{array}$$

your *annual gas/oil expenditure* would be $2,760.

Category	Annual Expenditure
Food, prepared at home	$_____ per year
Food, away from home	$_____ per year
Alcoholic beverages	$_____ per year
Clothing	$_____ per year
Gas and motor oil	$_____ per year

The chart below shows average annual expenditures in the United States by age, geographic region, and consumer unit. Circle the group that you belong in for each of the three categories: *age, region,* and *consumer unit.* Then answer the questions that follow.

Category	Food, Home	Food, Away	Alcoholic Beverages	Clothing	Gas and Motor Oil
Age					
Under 25	$2,330	$2,117	$448	$1,351	$1,974
25–34	3,393	2,836	491	1,965	2,754
35–44	4,509	3,340	462	2,235	3,347
45–54	4,452	3,244	505	2,228	3,298
55–64	3,710	2,646	525	1,622	2,818
65–74	3,421	1,917	343	1,381	2,045
75 and over	2,667	1,268	144	755	1,173
Region					
Northeast	$3,595	$2,824	$508	$2,068	$2,080
Midwest	3,252	2,541	501	1,866	2,408
South	3,311	2,470	382	1,692	2,522
West	3,822	2,988	488	2,042	2,389
Consumer Unit					
Buying for one person	$1,814	$1,514	$428	$ 971	$1,276
Buying for five or more people	5,564	3,656	353	2,719	3,531

Source: U.S. Census Bureau

1. Among people in your *age group*, do you spend more or less than the average on *food away from home?*

2. Among people in your *geographic region*, do you spend more or less than the average on *gas and oil?*

3. Write a statement comparing your *clothing* expenditures to the average in your *age group.* _____

_____

4. Find a category in which your expenditures are far above or below the average in your age, region, or consumer unit size. Why do you think you are not close to the average? Write a short explanation.

_____

Using Data to Write a Report

Businesses are very interested in information about how people spend their money. They use this information to make decisions on what to make, where to sell, and whom to market to. For example, by looking at data, an automobile maker can decide what kind of person is most likely to buy an inexpensive model and what kind of person would be interested in a fancy sports car.

Imagine you have a job in the business world and have put together the information in the chart below. It shows how much money people in different age groups spend on certain products each year, on average. It also shows how spending differs in four regions of the country. Take some time to look at the data, and then answer the questions that follow.

Annual Expenditures in the U.S. for Selected Products						
Category	Small Appliances	Non-alcoholic Beverages	Laundry and Cleaning Supplies	Clothing	Pets, Toys, Hobbies, and Playground Equipment	Reading Materials
Age						
Under 25	$ 49	$246	$ 83	$1,351	$380	$ 48
25–34	85	310	150	1,965	696	79
35–44	97	431	178	2,235	992	102
45–54	130	424	166	2,228	787	124
55–64	161	321	153	1,622	728	157
65–74	133	285	135	1,381	574	152
75 and over	96	207	91	755	266	132
Region						
Northeast	n/a	$333	$121	$2,068	n/a	135
Midwest	n/a	313	145	1,866	n/a	126
South	n/a	325	152	1,692	n/a	89
West	n/a	366	132	2,042	n/a	140

Source: U.S. Census Bureau

1. Suppose you work for a company that manufactures small appliances such as toasters and blenders. Your supervisor has made the decision to advertise to young people under the age of 35. Based on the data in the chart, you think she is making a mistake. Write a brief report stating what age group(s) spends the most in this category and why you think the company should not market to a younger age group. Be sure to use specific numbers in your report.

   _____

   _____

   _____

   _____

2. Now suppose you need to decide the region and age group to sell books to. Write a brief report giving your opinion. Use specific data from the chart to support your ideas.

   _____

   _____

   _____

   _____

3. In the space below, create a bar graph that shows clothing expenditures data among the different age groups given. Your horizontal axis should show age groups, and the vertical axis should show a range of 0 to $2,500, with increments of $500.

Using Graphs to Analyze Sales

When you work in the sales department of a company, you need to look at all kinds of data to answer different questions. What products sell well? How do sales among different products compare? Do sales change over time? Why do sales figures change?

Look at the graph below, and use the skills you have learned in earlier lessons to answer the questions that follow.

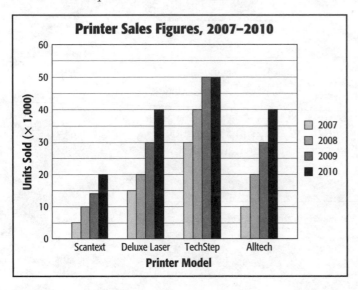

How many more printers did TechStep sell in 2010 than Alltech sold the same year?

STEP 1   Find the correct companies on the horizontal axis and the correct year on the key.

The 2010 TechStep bar rises to 50 on the vertical axis; Alltech rises to 40.

STEP 2   Be sure to read the scale on the vertical axis before you do your calculations.

The scale says x 1,000, which means you need to multiply 50 by 1,000 and 40 by 1,000.

TechStep sold 50,000 printers in 2010 and Alltech sold 40,000.

STEP 3   Do the necessary calculation:

50,000 − 40,000 = 10,000

TechStep sold **10,000** more printers than Alltech in 2010.

Use the sales chart on page 152 to answer the following questions.

1. About how many printers did Scantext sell between 2007 and 2010?

2. Which company showed the greatest growth in sales between 2007 and 2010?

3. Which company showed the least growth in sales between 2007 and 2010?

4. By what percent did Deluxe Laser sales increase between 2009 and 2010?

5. Your supervisor prefers to see data in line graph form, with years along the horizontal axis and sales on the vertical axis. Create a line graph using the bar graph data. The graph has been started for you below.

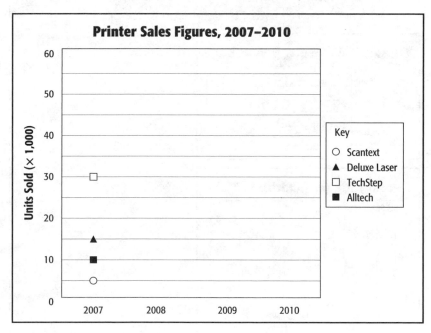

Writing a Newspaper Article

In this activity, you will write articles and letters to the editor like those you see in a newspaper. Try to use as many of the techniques from the Analyzing Data chapter as you can, including

summarizing
estimating
using correct information
changing the form of data
seeing trends
making predictions

Also, try to use information from two or more of the data sources below in your article.

U.S. ENERGY SUMMARY 2009

The United States increased its energy consumption by 28% between 1980 and 2009; in contrast, energy production was up only 6% in the same time period.

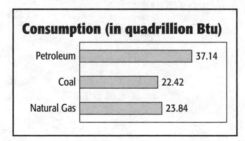

Consumption (in quadrillion Btu)

Petroleum — 37.14
Coal — 22.42
Natural Gas — 23.84

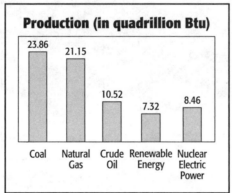

Production (in quadrillion Btu)

Coal — 23.86
Natural Gas — 21.15
Crude Oil — 10.52
Renewable Energy — 7.32
Nuclear Electric Power — 8.46

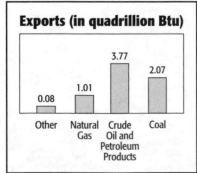

Exports (in quadrillion Btu)

Other — 0.08
Natural Gas — 1.01
Crude Oil and Petroleum Products — 3.77
Coal — 2.07

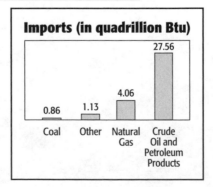

Imports (in quadrillion Btu)

Coal — 0.86
Other — 1.13
Natural Gas — 4.06
Crude Oil and Petroleum Products — 27.56

Source: Energy Information Administration, U.S. Department of Energy

1. Use one of the following statements for the first sentence in an article about our national energy situation. Write the article using any of the data from the previous page.

   a. The 2009 data concerning United States energy production tells an interesting tale about its priorities.

   b. The United States consumed over 37 quadrillion Btu of petroleum in 2009.

   c. The United States imported more than seven times the amount of crude oil and petroleum products it exported in 2009.

   _____

   _____

   _____

   _____

   _____

   _____

   _____

   _____

   _____

   _____

   _____

2. Do you have an opinion about our country's energy usage? For example, some people think that we are using up too much of our natural resources and that we should use more nuclear energy or renewable energy sources, such as solar or wind power. Other people believe that nuclear power is too dangerous and should not be used at all. Still other people believe that, no matter what our energy source is, we use too much of it on transportation, such as cars.

   On a separate sheet of paper, write a letter to the editor of a newspaper stating your opinion on energy in the United States. Use statistics to back up your point of view.

Using Data in Customer Service

A customer service representative at an electricity company helps people read and understand the bills they get in the mail. When a customer calls with a question or complaint, the representative looks at the data from the person's account and provides accurate and helpful information.

The chart below comes from the EnergyStar account of Stephen Marks. Take a look at the data and see if you can answer the question that follows.

ENERGY STAR Electric Bill Comparison			
	Current Month	**Last Month**	**Same Month Last Year**
Electric charges	$210.57	$178.81	$264.70
Total electricity use (kWh)	1,284	1,074	1,297
Delivery charges (per kWh)	7.6 cents	7.7 cents	7.7 cents
Delivery charges total	$97.58	$84.44	$99.89
Generation charges (per kWh)	8.8 cents	8.8 cents	12.7 cents
Generation total	$112.99	$95.37	$164.81
Average daily electric use (kWh)	40.1	35.8	40.5
Average daily temperature	67	57	70

ENERGYSTAR Electric Bill Comparison

- Mr. Marks calls and says he thinks there is a mistake on his bill this month. He is being charged more than last month. You explain to him that charges are based on total kWh (kilowatt hours) of electricity generated and delivered in the month. Can you explain to Mr. Marks the math EnergyStar uses to determine how much to charge?

STEP 1 In the current month column, find the number of kWh used and multiply it by the delivery charge per kWh.

1,284 × 7.6 cents (.076) = $97.58

STEP 2 Now multiply kWh used by the generation charge per kWh.

1,284 × 8.8 cents (.088) = $112.99

STEP 3 Add the total delivery charge and the generation charge together to find the total amount owed.

$97.58 + $112.99 = **$210.57**

Imagine that another customer calls you and would like help understanding her electric bill. Use the same steps you saw on page 156 to fill in the blanks on the bill below. Then answer the questions that follow.

ENERGY STAR Electric Bill Comparison			
	Current Month	**Last Month**	**Same Month Last Year**
Electric charges	$_____	$_____	$_____
Total electricity use (kWh)	985	1100	956
Delivery charges (per kWh)	7.6 cents	7.7 cents	7.7 cents
Delivery charges total	$_____	$_____	$_____
Generation charges (per kWh)	8.8 cents	8.8 cents	12.7 cents
Generation total	$_____	$_____	$_____
Average daily electric use (kWh)	28.9	31.8	26.5
Average daily temperature	69	60	68

1. The customer would like to know whether she is using more or less electricity this month than last month. What should you tell her, and what specific data should you reference?

   _____

   _____

2. The customer would like to know if per kilowatt hour delivery charges have changed since last year. She would also like to know if they have gone up or down and by how much. How should you respond?

   _____

   _____

3. Use the average daily temperature data to explain to the customer why her average daily electric use changes. Is there a correlation?

   _____

   _____

   _____

A Look at Political Polling in History

If you follow the news during election time in the United States, you are sure to be bombarded with statistics. For weeks before the election, you are showered with figures such as these:

- Candidate A is ahead of candidate B by 10%.
- Candidate C's name is recognized by 55% of the voting population.
- Candidate D rose 5% in the polls following a persuasive speech.

And now, with our rapidly improving technology, we even get up-to-the-hour statistics *as the election is taking place.* In **exit polls,** surveyors ask voters whom they voted for as they leave the polling site, and these figures are broadcast on television.

Now that you have a good understanding of samples and surveys, you know where these numbers come from. How reliable are these figures? The statistics you get from the major *independent* polling organizations (see pages 101–102) are generally very reliable. However, this was not always the case. Let's take a look at a famous case in which a survey provided very *unreliable* results.

President Thomas E. Dewey?

Polling and sampling techniques have improved greatly over the years. In the past, pollsters relied on a method of matching the U. S. Census exactly to get a representative sample. For example, if the census reported that 12% of the population was African American, pollsters made sure that 12% of their sample was African American. If 23% of the population commuted into a city to work, the sample had the same percentage.

Why, then, in 1948, did all three major national polls—Gallup, Roper, and Crossley—predict that Thomas Dewey would defeat incumbent Harry Truman? Truman won the election by a wide margin.

Many studies were done after this famous mishap. Looking back, we find that some mistakes were made in the surveying and sampling methods used by the polling organizations.

- The surveying took place weeks before the election. The surveys did not take into account a late trend favoring Truman.

- Pollsters were allowed to survey anyone they found in one area, while other areas and people weren't represented at all.

Since this embarrassing event, a great deal has been done to improve polling techniques. Surveys are now conducted right up to the day of an election. In addition, when surveyors are sent out to take a poll, they are given specific names and addresses of people to interview. These names are gathered by a true method of random sampling.

Basically, by putting the names of *every single American of voting age* on a list and choosing 1,200 of them at random, polling organizations ensure the principle of random sampling: *each member of the population is equally likely to be chosen.*

1. **Critical Thinking** Very early in an election year, citizens find out who the front-runners are in a political race—based on public opinion polls. Even before we hear all of a candidate's points of view, we read about his or her chances of winning. Do you think public opinion polls have a good or bad influence on our political system?

Consider This Situation

A voter does some research on all of the candidates for office and decides he likes how Candidate X stands on almost all issues. The voter then learns—from early public opinion polls—that Candidate X is favored by only 10% of the population. Believing that his candidate does not stand a chance of winning, the voter decides to cast his vote for Candidate Y. He does this because he is strongly against Candidate Z and he believes that Candidate Y can defeat Candidate Z.

2. Do you think that early public opinion polls influence voters? Do they influence the outcome of an election? Do they serve a legitimate purpose in a democracy, or do they cloud the issues? Can you come up with arguments for and against such polls?

Numbers in Advertising

"4 out of 5 doctors surveyed recommend brand X!"

"Now a hot dog that's 80% fat-free!"

"Get your laundry up to 30% cleaner!"

Have you ever noticed how many advertisements use statistical claims?

Are these numbers facts, and should they persuade us to buy these products?

The skills you have been working on throughout this book also apply to analyzing data found in advertisements and commercials. For each of the claims above, let's look at the numbers and *what they really mean.*

"4 out of 5 doctors surveyed recommend brand X!"

Ask yourself these questions.

- Can you tell how large a survey was conducted?
- Does the advertisement tell you what question was asked of the doctors in the survey?

The answers should make you question the claims being made.

Few, if any, advertisements will tell how the survey was conducted. Certainly, the makers of brand X would like you to believe that they surveyed thousands of doctors in a random sample and asked them what their favorite brand of pain reliever was. More likely, however, here is how they got their numbers:

They probably surveyed a small number of doctors, possibly doctors who work in their own research labs.

They most likely asked the doctors, "Would you recommend brand X pain reliever to your patients?" The doctors respond yes, but the advertisement does not say that they recommend ten other brands as well.

In fact, there is nothing *untruthful* in the advertisement. However, don't be fooled into assuming too much. When you see an advertisement like this one, think of not only what is being said but also what *isn't* being said. Remember what you have learned about source and bias!

"Now a hot dog that's 80% fat-free!"

Many people see a label that says "80% fat-free" and think that a product is low in fat. But what it's really saying is that the product is 20% fat. That's $\frac{1}{5}$ of what you're about to eat.

Whenever you see a percent on a label or in an advertisement, try to restate the data so it's more specific and more meaningful.

"Get your laundry up to 30% cleaner!"

Ask yourself, "30% cleaner than *what*?"

By not identifying what a product is being compared to, and by using impressive statistics, advertisements like this one can be misleading. Consumer beware!

---

1. **Critical Thinking** What is a good way to decide what kind of medication to buy? Can advertisements really help you make a decision?

   Choose three or four advertisements that use statistics to help sell a product or service. They can be television commercials or newspaper or magazine ads.

   a. Write the exact statement that uses statistics.

   b. Then write one or two reasons why the statistics do *not* help you decide whether to buy the product advertised.

Using Graphs to Track Productivity

Presenting data in graph form can be a powerful way to tell a story. Have you ever heard the expression "A picture is worth a thousand words?" Read the following excerpt from a business report and see if you can find the main point. Then look at the graph below the passage and see how the data "comes alive."

On Monday, 345 pieces were engraved, and there were 50 rejects. On Tuesday, there were 55 rejects, whereas on Wednesday there were 75 rejects. These numbers compare to 320 completed engravings on Tuesday and 300 completions on Wednesday....

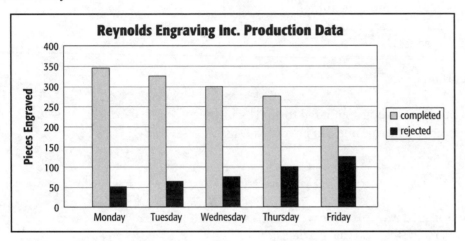

By looking at the data in graph form, you can easily make comparisons of productivity on different days of the week. You can also spot trends more easily.

Imagine that you work for Reynolds Engraving, and you are looking for ways to increase **productivity**—that is, to increase the number of usable pieces and reduce the number of rejects. You look at the data on the graph, and you can easily make the following statement. Fill in the blanks below:

As the week goes on, the number of completed pieces _____ and the number of rejected pieces _____.

Did you put **decreases** in the first blank and **increases** in the second blank?

Use the same graph to answer the questions that follow.

1. Your supervisor asks you which day of the week is most productive. Which statement is true?

   a. Tuesday is the most productive day of the week because completed pieces and rejects total almost 350.

   b. Wednesday is the most productive day of the week because completed pieces and rejects add up to 375.

   c. Monday is the most productive day of the week because it produced the most completed pieces and the fewest rejects.

2. Your supervisor tells you that there are different engraving workers each day of the week. He asks you to recommend retraining of some of the Reynolds engravers. In the space below, write a report with your recommendations, using specific days and specific data in your report.

Posttest B

This test has a multiple-choice format much like the GED and other standardized tests. Take your time and work each problem carefully. Circle the correct answer to each problem. When you finish, check your answers at the back of the book.

For problems 1–4, refer to the following graphs.

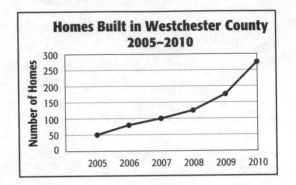

1. How many homes were built in Westchester County in 2007?

   a. 150   **c.** 100

   b. 125   **d.** 75

2. Which statement best summarizes the data of both graphs?

   a. The average size of homes built in Westchester County increased from 2005 to 2010.

   b. The number of homes built in Westchester County increased between 2005 and 2010.

   c. The number of homes in Westchester County decreased between 2005 and 2010.

   d. As the number of homes built in Westchester County increased between 2005 and 2010, the average size of the homes built increased.

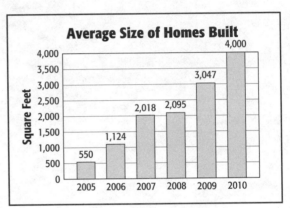

3. Which statement best summarizes homes built in Westchester County in 2005?

   a. There were 50 homes built in Westchester County in 2005.

   b. There were fewer homes built in Westchester County in 2005 than in 2006.

   c. There were 50 homes built in Westchester County in 2005, with an average size of 550 square feet per home.

   d. The size of the average home in Westchester County in 2005 was 550 square feet.

4. Use data from both graphs to *estimate* how many square feet of homes were built in Westchester County in 2008.

   a. 6,000   **b.** 37,000   **c.** 37,500   **d.** 60,500   **e.** 262,500

For problems 5–6, refer to the following data.

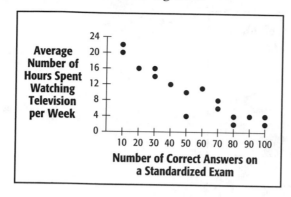

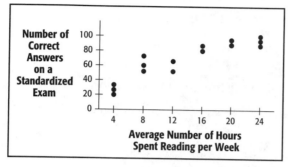

5. Which of the following statements is true based on the data?

   a. There is a positive correlation between the amount of television watched and the number of correct answers on a standardized exam.

   b. There is no correlation between hours of television watched and correct answers on a standardized test.

   c. As the number of hours spent watching television per week increases, the number of correct answers on a standardized exam decreases.

   d. The greater the number of hours spent watching television, the higher the standardized test score.

6. Which of the following statements is true based on the data?

   a. There is a positive correlation between the number of hours spent reading and the number of correct answers on a standardized exam.

   b. There is no correlation between hours spent reading and correct answers on a standardized test.

   c. As the number of hours spent reading increases, the number of correct answers on a standardized exam decreases.

   d. The more hours spent watching television, the fewer hours spent reading.

For problems 7–10, refer to the following data.

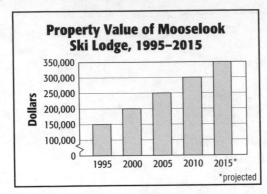

Property Value of Mooselook Ski Lodge, 1995–2015

*projected

7. Which of the following statements is true according to the graph?

   a. The property value of Mooselook Ski Lodge increased by $100,000 every 5 years between 1995 and 2015.

   b. The property value of Mooselook Ski Lodge will exceed $400,000 in 2020.

   c. The property value of Mooselook Ski Lodge rose steadily between 1995 and 2010.

   d. The property value of Mooselook Ski Lodge decreased by $100,000 every 5 years from 1995 and 2015.

   e. The property value of Mooselook Ski Lodge decreased by $50,000 every 5 years between 1995 and 2015.

8. If the trend continues, what will the property value of Mooselook Ski Lodge be in 2020?

   a. $300,000   b. $400,000   c. $450,000   d. $500,000   e. $550,000

9. Which of the following statements is true based on the graph?

   a. The property value of Mooselook Ski Lodge in 2010 was twice as much as its value in 2000.

   b. The property value of Mooselook Ski Lodge in 2010 was half its value in 2000.

   c. The property value of Mooselook Ski Lodge in 2015 is expected to be more than twice its 1995 value.

   d. The property value of Mooselook Ski Lodge in 1995 was its lowest value ever.

   e. The property value of Mooselook Ski Lodge cannot be expected to exceed $400,000.

10. By what percent did the value of Mooselook Ski Lodge rise between 1995 and 2010?

   a. 100%   b. 75%   c. 50%   d. 25%   e. 10%

For problems 11–14, refer to the following data.

Projected North Dover High School PTO Budget, 2010	
Enrichment	41%
Field Trips	28%
Technology	22%
Guest Speakers	8%

Margin of error: ±3%

11. Given the margin of error, what is the range of percents of the projected budget for field trips in 2010?

   a. from 28% to 31%

   b. from 39% to 45%

   c. from 40% to 42%

   d. from 25% to 31%

   e. from 19% to 25%

12. Considering the margin of error, what is the greatest fraction of the budget that will be spent on technology in 2010?

   a. $\frac{1}{4}$ b. $\frac{1}{2}$ c. $\frac{1}{3}$ d. $\frac{1}{5}$ e. $\frac{1}{8}$

13. If the North Dover PTO spends $15,000 in 2010, how much will be spent on guest speakers?

   a. $120   b. $700   c. $1,875   d. $100   e. $1,200

14. Given the margin of error, what percent of the projected North Dover High School PTO budget will be spent on enrichment?

   a. 38%

   b. 41%

   c. between 38% and 44%

   d. between 39% and 46%

   e. 22%

For problems 15–17, refer to the following data.

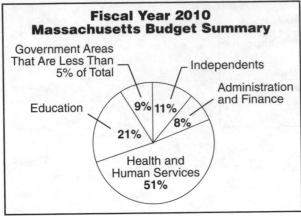

Source: mass.gov

15. Use data from the graph to *estimate* what the chances were that a dollar of the Massachusetts 2010 budget went to pay for education.

   a. $\dfrac{3}{4}$ **b.** $\dfrac{2}{3}$ **c.** $\dfrac{2}{5}$ **d.** $\dfrac{1}{5}$ **e.** $\dfrac{1}{10}$

16. Which of the following statements is true based on the graph?

   a. Approximately one dollar out of 10 was spent on health and human services in Massachusetts in 2010.

   b. Approximately five dollars out of 100 was spent on independents in Massachusetts in 2010.

   c. The largest percentage of the Massachusetts budget went toward education in 2010.

   d. Administration and finance expenses accounted for almost 1 dollar out of every 10 in the 2010 Massachusetts budget.

   e. The United States budgeted more of its 2010 dollars to health and human services than any other sector.

17. Approximately what fraction of the 2010 Massachusetts budget went to health and human services?

   a. $\dfrac{1}{51}$ **b.** $\dfrac{1}{20}$ **c.** $\dfrac{1}{10}$ **d.** $\dfrac{1}{2}$ **e.** $\dfrac{2}{3}$

For problems 18–20, refer to the following data.

J.W. Eaton Elementary School			
Grade	Number of Students Who Bring Their Lunch to School	Number of Students Who Buy Their Lunch	Total Number of Students
1	25	23	48
2	40	21	61
3	9	38	47
4	15	35	50
5	42	73	115

18. Which grade has the highest *rate* of students who bring their lunch to school?

   a. 1   **b.** 2   **c.** 3   **d.** 4   **e.** 5

19. According to the chart, about 60% of students at J. W. Eaton Elementary School buy their lunch. What number can you compare with this figure in order to decide if 60% is a large or a small number?

   a. the total number of students at J. W. Eaton Elementary School who bring their lunch

   b. the total number of students at J. W. Eaton Elementary School

   c. the percent of fifth graders who bring their lunch to school

   d. the percent of students who buy their lunch at a nearby elementary school

   e. the average cost of lunch at J. W. Eaton Elementary School

20. Which of the following statements is true at the J. W. Eaton School?

   a. More fourth graders bring their lunch than buy their lunch.

   b. Almost $\frac{3}{4}$ of first graders buy their lunch.

   c. There are 50 students in the third grade.

   d. More than 50% of second graders buy their lunch.

   e. Of the fourth graders, 70% buy their lunch.

For problems 21–24, refer to the following data.

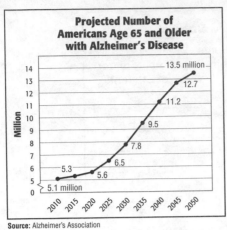

Projected Number of Americans Age 65 and Older with Alzheimer's Disease

Source: Alzheimer's Association

21. By how many millions is the number of Americans 65 and older with Alzheimer's disease expected to increase between 2025 and 2035?

   a. 0.3   **b.** 0.5   **c.** 3   **d.** 12   **e.** 30

22. What is the range of millions provided on this graph?

   a. 5.3 to 13.5   **b.** 5.1 to 13.5   **c.** 5.3 to 12.7   **d.** 5.1 to 12.7   **e.** 5.1 to 10

23. Which of the following statements best summarizes the graph?

   a. The graph shows the projected number of Americans age 65 and older with Alzheimer's disease between 2010 and 2050.

   b. The graph shows the growing number of people with Alzheimer's disease.

   c. The graph shows the population of Americans age 65 and older between 2010 and 2050.

   d. The graph shows the huge increase of Alzheimer's patients in the United States over several years.

   e. The graph compares rates of Alzheimer's disease between 2010 and 2050.

24. According to the graph, which statement is true?

   a. The number of Americans age 65 and older with Alzheimer's disease is projected to more than double between 2015 and 2035.

   b. More than 13 million Americans age 65 and older will die from Alzheimer's disease in 2050.

   c. The number of Americans age 65 and older with Alzheimer's disease is projected to more than double between 2010 and 2040.

   d. The number of Americans age 65 and older with Alzheimer's disease is projected to increase between 2010 and 2050 due to increased longevity in the population.

   e. The number of Americans age 65 and older with Alzheimer's disease is projected to increase by 3 million between 2010 and 2015.

For problems 25–28, refer to the following survey.

A recent survey was conducted regarding eyeglasses and contact lenses wear. The results showed that 57% of visually impaired Americans own both contact lenses and glasses, 33% own only contact lenses, and 10% own only glasses. It was also found that 75% of those who own both contact lenses and glasses prefer to wear their contact lenses to their glasses.

25. Which of the following companies would most likely benefit from the statistics in the survey above?

   a. Tierney Eye Glass Wear, Inc.

   b. See Perfect Contact Lenses Manufacturers

   c. The Institute of Ophthalmology

   d. Pediatric Eye Associates

   e. Northeast Eye World

26. Which of the following questions might the surveyors ask to ensure that Americans who own both eyeglasses and contact lenses say they prefer wearing contact lenses to glasses?

   a. Which do you wear more often, contact lenses or glasses?

   b. Would you prefer to wear soft and comfortable contact lenses or your glasses?

   c. Which do you prefer to wear—glasses or contact lenses?

   d. Are your eyeglasses convenient to wear?

   e. Do you occasionally have trouble putting in your contact lenses?

27. Which of the following would be a biased sample on which to conduct the survey?

   a. 500 employees at a bank

   b. 1,000 people shopping at a mall

   c. 1,000 people walking out of a contact lenses fitting office

   d. 100 patients at an eye doctor's office

   e. 100 people at a county fair

28. According to the survey, what are the chances that an American wears only glasses?

   a. 1 out of 10   **d.** 1 out of 4

   b. 1 out of 5   **e.** 1 out of 20

   c. 1 out of 3

For problems 29–31, refer to the following data.

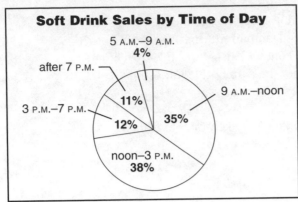

Source: Best Beverage, Inc.

29. Which time of day shows the largest sales percentage?

    a. 9 A.M. – noon

    b. noon – 3 P.M.

    c. 3 P.M. – 7 P.M.

    d. after 7 P.M.

    e. 5 A.M. – 9 A.M.

30. What is the purpose of the graph above?

    a. to show how many soft drinks were sold by Best Beverage in a single day

    b. to show total sales in dollars of Best Beverage soft drinks

    c. to show the percent of time spent selling soft drinks during a day

    d. to show the increase in soft drink sales between 5 A.M. and 7 P.M.

    e. to show the percent of Best Beverage soft drink sales during different time periods of the day

31. Which of the following statements is false?

    a. 23% of soft drink sales occur from 3 P.M. until after 7 P.M.

    b. Most soft drink sales occur at lunch time—between noon and 3 P.M.

    c. Less than 5% of soft drink sales occur between 5 A.M. and 9 A.M.

    d. Percent of soft drink sales range from 4% to 38% over the course of the day.

    e. 12% of soft drink sales occur after 7 P.M.

For problems 32–35, refer to the following data.

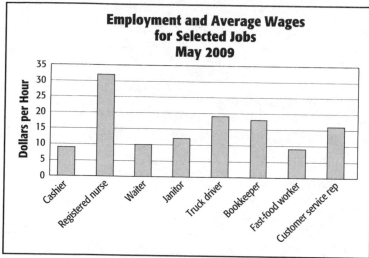

Source: Bureau of Labor Statistics

32. How many dollars per hour did a customer service representative earn in May 2009?

   a. just over 30

   b. about 20

   c. just over 15

   d. slightly under 15

   e. slightly more than 10

33. Which of the following statements summarizes the data?

   a. The graph shows the average hourly wages of selected jobs in May 2009.

   b. The graph shows how much more a registered nurse earns than any other profession.

   c. The graph shows how many people were employed at different jobs in May 2009.

   d. The graph shows the average salary of a worker in the United States in May 2009.

   e. The graph shows the rise in hourly wages in May 2009.

34. According to the chart, which of the following statements is true?

   a. A registered nurse is the top wage earner in the United States.

   b. Of the jobs listed, cashiers and fast food workers are the lowest hourly wage earners.

   c. Of the jobs listed, waiters are the lowest hourly wage earners.

   d. In May 2009, a truck driver earned more than twice what a waiter earned per hour.

   e. In May 2009, a janitor earned an average of $8.50 per hour.

35. According to the graph, which job had the highest average hourly wage in May 2009?

   a. truck driver

   b. customer service rep

   c. waiter

   d. registered nurse

   e. bookkeeper

POSTTEST B CHART

If you missed more than one problem on any group below, review the practice pages for those problems. Then redo the problems you got wrong. If you had a passing score, redo any problem you missed.

Problem Numbers	Skill Area	Practice Pages
3, 7, 16, 20, 23, 31	making statements about data	11–12
19	comparing numbers	16–18
18, 34	numbers vs. rates	22–25
22	scale and range	30–33
1, 3, 21, 24	line graphs	36–40
32, 35	visual statistics	41
17, 29, 30	circle graphs	44–46
13, 33	using only the information given	66–68
4, 15	estimation	71–73
2	using more than one data source	74–76
8, 9	seeing trends/making predictions	82–85
15, 28	using data to find probability	90–92
27	random samples	96–98
11, 12, 14	margin of error	103–105
10	zero on a scale	116–118
5, 6	understanding correlation	119–123
25, 26	understanding bias	124–129

ANSWER KEY

1. **b.** Tokyo. 37,000,000 is the largest number on the chart.

2. The table above shows the **projected 2020 population** to the nearest million for five cities.

3. Answers may vary. A sample statement is given below.

   The expected 2020 population of Dhaka will be about 2,000,000 higher than the expected New York City population.

4. **c.** 11,000,000

   $$\begin{array}{r} 37{,}000{,}000 \quad \text{(Tokyo)} \\ -\ 26{,}000{,}000 \quad \text{(Mumbai)} \\ \hline 11{,}000{,}000 \end{array}$$

5. **a.** 40%

6. being caught in traffic

7. Answers may vary. A sample statement is given below.

   About 50% of the people surveyed get angry because of slow restaurant service.

8. 48%

   65% + 65% + 50% + 40% + 20% = 240%

   240 ÷ 5 = 48%

9. **c.** 77%

10. **a.** twenty

    1998: about 60%

    2008: about 80%

    80 − 60 = 20

11. **b.** more than 80%

    The graph shows the percent rising, so the percent in 2010 will probably be greater than the 80% in 2008.

12. **c.** There were fewer automobile deaths among youths 16–24 in 2008 than in prior years.

    You cannot assume that there are fewer deaths when seatbelt use rises.

13.

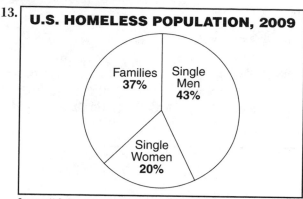

Source: U.S. Department of Housing and Urban Development

14. **c.** 240

    20% of 1,200 is the same as 20% × 1,200, or 240.

15. Answers may vary. A sample statement is given below.

    More than twice as many single men as single women were homeless in the United States in 2009.

16. **c.** 3 out of 5

    $\dfrac{3}{5}$ (2, 6, and 10 are multiples of 2)

    (5 cards in all)

17. **a.** 0

    Since there are no fives, the probability of choosing a five is 0.

18.

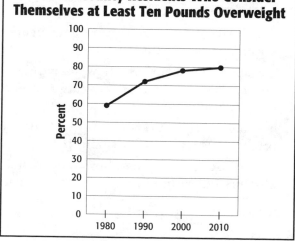

19. Answers may vary. A sample statement is given below.

    In general, as the age of the automobile increases, the number of repairs per year increases.

Page 10

Answers may vary.

Page 12

Answers may vary.

Pages 14–15

Yes	No	No Opinion
IIII	II	I

1. 95%, $\frac{95}{100}$; 95:100 or 19:20

2. 8%; $\frac{8}{100}$; 8:100 or 2:25

3. 70%; $\frac{7}{10}$; 7:10 or 70:100

4. 30%; $\frac{3}{10}$; 60:200 or 30:100 or 3:10

5. 1%; $\frac{1}{100}$; 1:100

6. 80%; $\frac{4}{5}$; 4:5 or 80:100

Pages 17–18

Answers may vary. Sample statements are given below.

1. $80,000 for the school lunch program would be a lot of money if **the total school budget was $100,000.**

2. $80,000 for the school lunch program would be a small amount of money if **$150,000 had been allotted for the program last year.**

Answers may vary. Here are some examples.

3. It would be helpful to know **how many people had been *expected* to attend.**

4. It would be helpful to know **how many people the company employs.**

5. It would be helpful to know **how many guns are confiscated in this area during a certain length of time.**

6. It would be helpful to know **how much money the artist usually gets for a painting.**

7. It would be helpful to know **how many years old the car is or how many miles it has on it.**

Pages 19 and 21

1. nine hundred thousand

2. eight million

3. ninety million

4. 100,000,000

5. 2,000,000,000

6. 700,000

7. 1,000

8. 10

9. 10

10. 100

11. 10

Pages 23–25

Highway

B	0.001	=	0.1 %
D	0.0033	=	0.33 %
G	0.0015	=	0.15 %
HH	0.0015	=	0.15 %
MM	0.002125	=	0.21 %

Answers may vary. Sample statements are given below.

1. Kramer, Inc., had the least number of satisfied employees, 13.

2. Psytech had the greatest number of satisfied employees.

3. Psytech, with 11.5%, had the lowest satisfaction rate.

4. HaML & Co. had the highest satisfaction rate, almost 25%.

5. Although Psytech had the highest number of satisfied employees, its employee satisfaction rate was the lowest.

6. The unemployment rate decreased from 5.1% in 2005 to 4.6% in 2006.

7. Blacks have a higher unemployment rate than the total work force.

8. In 2009 the black unemployment rate was about 5% greater than that of whites.

Pages 27–29

1. 1

2. 74

3. 152

4. T

5. T

6. T

Answers may vary. Sample statements are given below.

7. A one-ounce serving of peanuts **contains four more grams of fat than one ounce of potato chips.**

8. Although corn chips have more calories than potato chips, they contain less fat.

9. Answers may vary.

10.

Percent of Americans Holding Credit Cards			
	$50,000 or More	$25,000–$34,999	Less than $15,000
Bank Credit Card	94%	73%	36%
Gasoline Credit Card	54%	33%	19%
Express Gold Card	14%	3%	1%

Source: Money Express magazine

11. 73

12. gasoline

13. $25,000–$34,999

14. 40

15. F

16. T

17. F

18. T

Answers may vary. Sample statements are given below.

19. The bank credit card **had the highest percent of credit card holders in every category.**

20. Americans earning less than $15,000 had the lowest percent of credit card holders.

Pages 31–33

1. 90%

2. 20%

3. Answers may vary.

4. Park and Recreation Land by County

5. counties

6. acres (in hundreds)

7. 0 to 20

8. 4

Answers may vary. Sample statements are given below.

9. Dawton County has more park and recreation land than **Kane County.**

10. Kane County **has less park and recreation land than the other counties shown.**

11. San Pedro County has the most park and recreation land of all the counties shown.

12.

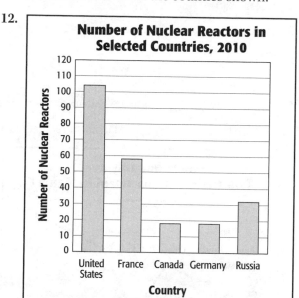

Source: International Atomic Energy Agency

Answers may vary. Sample statements are given below.

13. In 2010 the United States had **more nuclear reactors than any of the other countries shown.**

14. Canada had **the same number of nuclear reactors as Germany in 2010.**

15. France had **more nuclear reactors than Russia in 2010.**

Page 35

1. F

2. F

3. T

4. T

Page 37

1. 30 years old

2. 5 years

3. above

Answers may vary. Sample statements are given below.

4. The median U.S. age was the same in 1980 as it was in **1950.**

5. The median age in the United States rose between 1820 and 2010.

6. The median age in the United States in 1990 was about 33 years old.

7. Answers will vary.

1. risen

2. rise

3. A12

4. C16

5. Answers will vary. A sample statement is below.

People preferred less expensive alarms starting in 2010.

6.

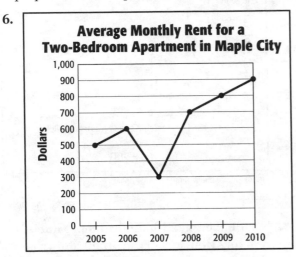

7. A line graph shows the change over time better than a bar graph.

Answers may vary. Sample statements are given below.

8. Rental prices in Maple City were about $600 higher in 2010 than in 2007.

9. Except for a sharp drop in 2007, rental prices rose between 2005 and 2010.

Page 41

Answers may vary. Sample statements are given below.

1. Between 1940 and 2008, the number of farms in the U.S. declined.

2. Between 2000 and 2008, the number of farms in the U.S. stayed the same.

Pages 42–43

1. 18

2. Australia

3. 2

4. 21

(**Note:** If you chose a different value in your key, your graph will be different from this one. Be sure that you have included the proper number of symbols on your graph.)

Average Annual Salary for a Full-Time Worker JT Manning, Inc.

Key:
$ = $10,000

'05 '06 '07 '08 '09 '10

5–6. Answers may vary.

Pages 45–46

Answers may vary. Sample statements are given below.

1. There are fewer students from **the Middle East than from Asia.**

2. Students from Asia accounted for more than half of foreign students.

3. National defense receives about twice as much as states and localities.

4.

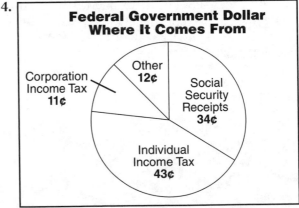

Source: Office of Management and Budget

Answers may vary. Sample statements are given below.

5. More than half of federal money **comes from income taxes.**

6. Corporate income taxes bring in almost as much as other smaller sources combined.

Pages 47 and 49

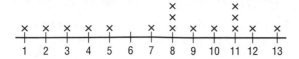

1.

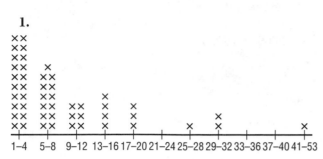

2. One representative is the least number from any state. Seven states have one representative.

3. California, with 53 representatives, has the greatest number. No other state has close to that many representatives.

4. There is a cluster of values between 1 and 12 representatives.

5. Answers may vary. Two sample statements are given below.

   a. Many states have **fairly small populations and therefore only 1 to 4 representatives.**

   b. California has the largest population, and Texas has the second largest.

6. Answers may vary.

Page 51

1.

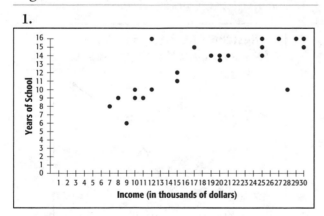

2. Yes, the higher the number of school years completed, the greater the salary.

Page 53

1. 4 miles   3. $32

2. $70

Page 54

1. $12.25

2. 6′10″

Page 55

1. 1,700:2   260:2
   1,650:1   250:9
   1,600:1   240:3
     290:2   230:1
     270:2   220:1

2. 250

3. $0.49

4. 10 weeks

Page 56

1. Mean: $646.67
   Median: $575
   Mode: $1,100

   Best representation of typical value: **mean or median**

2. Mean: 254 people
   Median: 200 people
   Mode: 225 people

   Best representation of typical value: **median or mode**

Page 57

1–2. Answers may vary.

Pages 61–62

1. No.
   The *what* is inaccurate; the graph shows **different occupations,** not necessarily the best-paying jobs.

2. No.
   The *where* is inaccurate; the graph shows the price of milk at **Crescent Farm Dairy.**

3. No.
   The *what* and *how* are inaccurate; the chart shows Americans who **believe** it's all right to lie **in percent,** not numbers.

4. Yes.

Pages 64–65

1. The statement does not take into account the label (× 100) on the vertical axis; the number of passengers carried is 1,200, not 12.

2. The statement does not include the necessary time: 2011.

3. The statement does not tell what the 600 refers to; Carling Coach carried 600 passengers in 2011.

Answers may vary. Sample statements are given below.

4. According to the County Taxi Association, Carling Coach carried **half as many** passengers as Yellow Cab in 2011.

5. In 2011 At Your Service cab company carried **200 fewer passengers than** Yellow Cab, according to the County Taxi Association.

6. Yellow Cab carried **the greatest number of** passengers in 2011.

7. Speedy Taxi Co. carried $\frac{1}{3}$ **the number of passengers** than Yellow Cab in 2011.

Page 68

1. Just under half of people responding to the survey felt that safety from abuse is important. This does not mean that many children were victims.

2. The graph shows *percent* of people interviewed, not actual numbers—46%, not 46 people.

3. The graph describes people in Massachusetts, not the entire country.

Page 70

1. more than

2. between

3. almost, approximately

Answers may vary. Sample statements are given below.

4. The average annual cost of car insurance will rise steadily between 2009 and 2012.

5. The average annual cost of car insurance in 2012 is expected to be $1,400.

Page 73

1. In 2020, Virginia will need **approximately 3,500** family physicians.

2. In 2006, Texas had **about two times** as many family physicians as Michigan had that year.

3. **Virginia and Michigan** will need about the same number of family physicians in 2020.

4. Of the states listed, **Florida** will need the largest increase in family physicians between 2006 and 2020.

5–6. Answers will vary.

Pages 75–76

1. Graph 1

2. Both

3. Both

4. Graph 2

5. **a.** Both
   b. $\frac{1}{10}$ $(\frac{140}{1,400} = \frac{1}{10})$

6. **a.** Both
   b. 42 (20% of 210 = 42)

7. **a.** Graph
   b. 210 (15% of 1,400 = 210)

8. **a.** Neither
   b. The chart and graph give no information about which sexes participated in which activities.

Pages 79 and 81

1. 2006: 31%; 2009: 49%; 2010: 56%

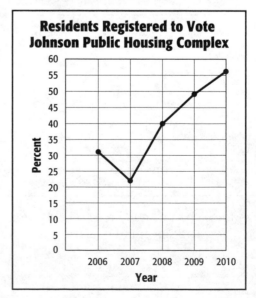

Residents Registered to Vote
Johnson Public Housing Complex

2. a. The *percent* of registered voters *rose* between 2009 and 2010, even though the first graph showed the *number* of registered voters falling.

   b. Answers may vary.

3. Step 1. 540

   Step 2. Medicare: $140 \div 540 =$ about 26%

              BC/BS: $120 \div 540 =$ about 22%

              Private: $200 \div 540 =$ about 37%

              No insurance: $80 \div 540 =$ about 15%

   Step 3.

Insurance	Medicare	Blue Cross/ Blue Shield	Private	No Insurance
Percent of Patients	26%	22%	37%	15%

4.

Sun County Homeless Population	
Downtown Area	37% or 3,515
Central Square	29% or 2,755
East Side	22% or 2,090
West Side	10% or 950
Outlying	2% or 190

Pages 84–85

1. $900

Answers may vary. Sample statements are given below.

2. If the trend continues, the approximate monthly rent for a two-bedroom/two-bathroom apartment in Del County in 2011 will be $1,000.

3. If the trend continues, the number of new SE County Union members will be close to 600 in 2012.

4. The number of new SE County Union members **rose steadily** between 2007 and 2011.

5. The number of new SE County Union members **was less in 2010** than the projected number for 2020.

6. There were **about 50 fewer** new SE County Union members in 2008 than in 2009.

Pages 86–87

1–3. Answers may vary. A sample statement is given below.
The total number of heads vs. the total number of tails is the same.

4. $\dfrac{1}{6}$

5. $\dfrac{3}{6}$ or $\dfrac{1}{2}$

6. $\dfrac{2}{6}$ or $\dfrac{1}{3}$

7. $\dfrac{9}{10}$

8. $\dfrac{1}{10}$

Pages 88–89

1. $\dfrac{1}{4}$ ◄—(number of fours) ◄—(total number of cards)

2. $\dfrac{3}{4}$ ◄—(4, 6, and 8 are greater than 3) ◄—(total number of cards)

3. $\dfrac{4}{4}$ or 1 (all numbers are divisible by 2)

4. 0 (none of the numbers is greater than 8)

5. Answers may vary.

6. 1

7. 1

8. Answers may vary.

9. 1

10. 0

11. 1

12. 1

13. 0

14. Answers may vary.

Answers may vary. Some examples are given below.

15. By the year 2020, animals will be able to talk as people do.
A 3-year-old will be elected president in the next election.
$1 + 1$ will no longer be equal to 2.

16. Someone in the world will make a new discovery this year.
A baby will be born in this country this year.
Some part of the earth will be in darkness today.

Pages 91–92

1. 30%

2. $\dfrac{30}{100} = \dfrac{3}{10}$

3. 3 in 10

4. **Step 1.** $1,280 + 960 + 320 = 2,560$

   Step 2. $\frac{1,280}{2,560} = \frac{1}{2}$

5. $\frac{5}{8}$ 7. less than $\frac{1}{2}$

6. $\frac{1}{8}$ 8. less than $\frac{1}{3}$

Page 95

1. Population: test scores of sixth-graders in state
   Sample: test scores of sixth-graders in six towns

2. Population: low-income people in city
   Sample: people in city low-income housing units

3. Population: defective canisters on early-morning shift
   Sample: defective canisters produced between 12:00 and 1:00 A.M.

4. Population: toy store's merchandise
   Sample: three toys from store

5. Population: all the baked goods in the store
   Sample: chocolate chunk cookies to passersby

Page 98

1. Population: Americans
   Sample: 200 people living near an army base
   The sample is not representative because people living near an army base may feel differently about defense than others.

2. Population: freshest food at the local markets
   Sample: chickens in five different stores
   The sample is not representative of all food in the markets. Where chicken was fresh, produce, for example, may not have been; or vice versa.

3. Population: people of the United States
   Sample: members of American Small Business Owners Association
   The sample is not representative of Americans because it excludes all who do not own a small business.

4. Population: defective circuit boards each month
   Sample: first 100 circuit boards one Tuesday morning
   The sample is not representative because the Tuesday morning shift might be the best or worst shift of the week.

Pages 100–102

Answers should include some of the responses listed below.

1. Mail or Internet
   Advantages: inexpensive; can reach a wide and varied sample
   Disadvantages: low response rate; responses are often not representative of those who received the questionnaire

2. Telephone
   Advantages: accurate results because of good participation (more representative of population); easy to use
   Disadvantages: leaves out people without telephones; hard to verify information

3. Personal Interview
   Advantages: reliable; accurate; good participation (more representative of population)
   Disadvantages: expensive; often survey fewer people

4. Answers may vary.

Pages 103–105

1. 20%

2. $20\% - 2\% = 18\%$
   $20\% + 2\% = 22\%$
   The range is from 18% to 22%.

3. The city council approval rating declined from March to April of 2010.

4. 60%

5. from 57% to 63%

6. 55%

7. from 52% to 58%

8. March: 57%
   April: 58%

9. Yes, when you consider the possible effect of the margin of error on the figures for both months.

10. from 67% to 75%

11. Yes, it is possible.
    $39\% + 4\% = $ a possible 43% who said they are happier
    $46\% - 4\% = $ a possible 42% who said they are not happier

39% ± 4% = 35% to 43% who said they are happier

46% ± 4% = 42% to 50% who said they are not happier

12. 525 teenagers (35% of 1,500 = 525)

13. 105 teenagers (7% of 1,500)
225 teenagers (15% of 1,500)
150 teenagers (10% of 1,500)

Page 107

1–3. Answers may vary.

Pages 110–111

Answers may vary. The following are sample statements.

1. a. Although oral contraceptives have a low failure rate in preventing pregnancy, this does not make them the *best* method of birth control. Other factors such as a woman's age, health, lifestyle, and preferences need to be considered, and what is best for one person is not necessarily best for everyone.

   b. Oral contraceptives have a failure rate of 6%, compared to a 16% failure rate for condoms and an 18% failure rate for diaphragms.

2. a. Although North County pays higher police salaries than the other four counties, the salaries might not be *too* high. Maybe North County requires more skills among its police force, or perhaps the other counties' police salaries are too low.

   b. Full-time North County police officers were paid an average of $30,000 per year—the highest salary among the five counties listed.

3. a. Fire deaths in Massachusetts are not necessarily caused by poor fire protection. More fatal fires may occur in Massachusetts because of careless smoking, more frequent use of chemicals, more flammable building materials, and so forth.

   b. In the state of Massachusetts, 29 percent of fire deaths are caused by careless smoking.

Page 115

1. a. Statement 1 refers to Graph A.

   b. Statement 2 refers to Graph B.

2. Answers may vary. A sample statement is given below.
According to Graph A, the population in Harris County was approximately 40,000, while in Drake County it was about 70,000.

3. a. Yes.

   b. No.

4. Answers may vary. A sample statement is given below.
In Graph A the scale increases by 10,000. In Graph B the scale increases by 50,000. Because the spacing between 0 and 7 on Graph A is wider than the spacing between 0 and 7 on Graph B, the bars on Graph A are longer.

Page 118

1. The statement is not accurate because there is no zero on the scale.

2.

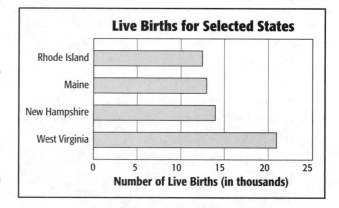

3.

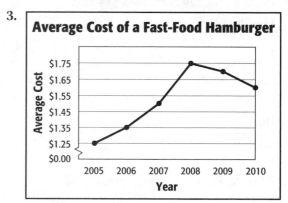

Pages 121 and 123

1. Answers may vary.

2. a. There **is** a correlation between **monthly take-home pay** and **number of television sets owned**.

b. The analysis is not accurate. Although there is a correlation, an increase in one value does not *cause* an increase in the other.

3. a. As the number of school years completed **rises,** the number of hours spent reading per week **rises.** (*Decreases* may also be used in both blanks.)

b. There is a **positive** correlation between school years completed and time spent reading.

Pages 125, and 127–129

1. Answers may vary.

2. Biased. The group Citizens for Smith has an interest in more people voting for Smith.

3. Biased. A soft-drink company would want more teenagers to prefer its brand.

4. Biased. A group called Americans Tough on Criminals would most likely want to find large numbers of people in favor of capital punishment.

5. Unbiased. A newspaper would probably have no preference concerning how people feel about gun control.

6. Biased. A person trying to get financial support would most likely want to find a large number of people interested in his or her idea.

7. Unbiased. A television network generally shows no preference for a particular candidate.

8. In general, answers are yes.

9. Biased. The question asked is loaded. It emphasizes the advantages of a new factory without mentioning the possible disadvantages.

10. Unbiased. The network is most likely looking for facts and has no preference as to how the numbers turn out.

11. Biased. This question points out the disadvantages without mentioning the advantages.

12. Biased. The sample surveyed is not representative of Americans in general. People shopping at an exclusive shopping mall are not likely to be affected by a recession in the same way as other Americans.

13. Biased. People leaving an ice cream shop are more likely to prefer ice cream than the general population.

Page 133

1. Answers may vary. A sample statement is given below.
Between January and July 2010, there were more homicides in April than in any other month.

2. d. 2010 homicides in Richmond, Virginia between January and July ranged from a low of zero to a high of nine.

Posttest A, Pages 134–143

1. service
sales: $26 \div 110 \approx 0.24$
service: $10 \div 25 = 0.4$

2. b. the percent of people who carpool to other nearby industries

3. Answers will vary. The following is a sample statement.
The maintenance department has 10 times as many employees who carpool to work as management does.

4. a. The graph shows the percent of the U.S. population between 1940 and 2010 who lived in multigenerational households.

5. from 0 to 30

6. Answers will vary. The following is a sample statement.
The percent of the U.S. population living in multigenerational households rose from 12.1% to 15.1% between 1980 and 2000.

7. c. In 1990, the percent of the U.S. population living in multigenerational households was actually about 14%, not 20%.

8. c. American Potato Growers, Inc.

9. a. Would you prefer a hot, buttered baked potato or a bowl of rice?

10. c. 3,000 participants at the Potato Lovers Picnic

11. 1 out of 5 or $\frac{1}{5}$

12. cold beverages in cans

13. c. to show percentage of vending machine sales for different categories

14. Answers will vary. The following is a sample statement.
19% of vending machine sales were for snack food.

15. $10(1.6 + 8.4)$

16. **a.** The graph shows hours on average that high school students spend engaged in different activities on weekdays.

17. Answers will vary. The following is a sample statement.
    Unemployed high school students spend more time on average socializing and relaxing than they do on sports and exercising.

18. **c.** Unemployed high school students spend 1.5 hours more on exercise and sports than employed high school students.

19. 13%

20. **b.** The percentage of whites in the U.S. is projected to decrease between 2005 and 2015, while the percentages of other racial groups will increase.

21. Answers will vary. The following is a sample statement.
    By 2015, whites will make up a smaller percentage of the U.S. population than in previous years.

22. Hispanics made up 13% of the population in 2005 and are projected to be 16% in 2015. Therefore, **15%** would be a good estimate of the Hispanic population in 2012.

23. Yes.
    As the number of leisure hours rises, the number of medicines taken also rises.

24. positive correlation
    As age increases, so does the number of medicines taken per year.

25. Answers will vary. The following is a sample summary.
    The number of medicines taken increases as age and the number of leisure hours increase. Of these three factors, age is probably the one that has an effect on the other two. Older people tend to be retired, thus having more leisure time. In addition, older people tend to have more medical problems (high blood pressure, arthritis, and so forth) that may require medication.

26. Answers will vary. The following is a sample statement.
    The average monthly Calereso commission is projected to be $1,900 in 2012.

27. **c.** The average monthly commission at Calereso rose steadily between 2009 and 2011.

28. The 2011 average monthly commission is *not* twice the amount earned in 2009. It appears that way on the graph because there is a zero break.

29. from 6% to 14%
    $10\% - 4\% = 6\%$
    $10\% + 4\% = 14\%$

30. Yes.
    $46\% + 4\% = 50\% = \frac{1}{2}$

31. 200,000 votes
    46% is approximately 50%
    50% of 400,000 = 200,000

32. Yes.
    If Crane actually received $46\% - 4\%$, or 42%, and Davidson received $39\% + 4\%$, or 43%, Davidson would be the winner.

33. $\frac{3}{5}$
    $60\% = \frac{60}{100} = \frac{3}{5}$

34. Answers will vary. The following is a sample statement.
    Hispanics in Hunnewell schools make up 5% of the school population, which is about half the Asian American population.

35. 620 students
    5% of 12,400 = $0.05 \times 12,400 = 620$

USING NUMBER POWER

A Look at the Census, Pages 146–147

1. 2000

2. 11.3 to 13.3

3. A homeless person (or people living in shelters) would less likely be counted in census data. Therefore, the numbers would probably be lower than the actual population.

Are You Average?, pages 148–149

1–4. Answers may vary.

Using Data to Write a Report, Pages 150–151

1–2. Answers may vary.

3.

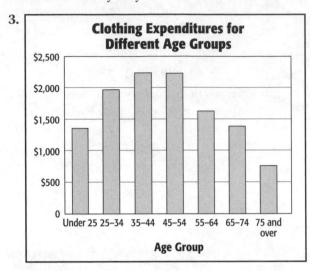

Using Graphs to Analyze Sales, Pages 152–153

1. 50,000

2. AllTech

3. ScanText

4. $33\frac{1}{3}\%$

5.

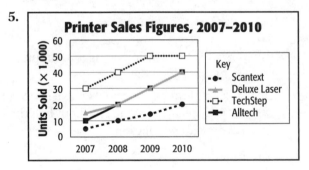

Writing a Newspaper Article, Pages 154–155

1–2. Answers will vary.

Using Data in Customer Service, Pages 156–157

ENERGY STAR Electric Bill Comparison			
	Current Month	Last Month	Same Month Last Year
Electric charges	$161.54	$181.50	$195.02
Total electricity use (kWh)	985	1100	956
Delivery charges (per kWh)	7.6 cents	7.7 cents	7.7 cents
Delivery charges total	$74.86	$84.70	$73.61
Generation charges (per kWh)	8.8 cents	8.8 cents	12.7 cents
Generation total	$86.68	$96.80	$121.41
Average daily electric use (kWh)	28.9	31.8	26.5
Average daily temperature	69	60	68

1. The customer is using less energy this month (985 kWh compared to 1,100).

2. Delivery charges have gone down 0.1 cent since last year. (7.7 − 7.6 = 0.1)

3. As the average daily temperature goes down, the number of kWh used goes up.

A Look at Political Polling in History, Pages 158–159

1–2. Answers will vary.

Numbers in Advertising, Pages 160–161

1. **a.** Statements will vary depending on the statistics used in the advertisements.

   b. Reasons will vary.

Using Graphs to Track Productivity, Pages 162–163

1. **c**

2. Answers will vary.

Posttest B, Pages 164–174

1. **c.** 100

2. **d.** As the number of homes built in Westchester County increased between 2005 and 2010, the average size of the homes built increased.

3. **c.** There were 50 homes built in Westchester County in 2005 with an average size of 550 square feet per home.

4. **e.** 262,500
   There were about 125 homes built in 2008, with an average size of about 2,100 square feet. 125 × 2,100 = 262,500

5. **c.** As the number of hours spent watching TV per week increases, the number of correct answers on a standardized test decreases.

6. **a.** There is a positive correlation between the number of hours spent reading and the number of correct answers on a standardized exam.

7. **c.** The property value of Mooselook Ski Lodge rose steadily between 1995 and 2010.

8. **b.** $400,000
   $350,000 + $50,000 = $400,000

9. **c.** The property value of Mooselook Ski Lodge in 2015 is expected to be more than twice its 1995 value.

10. **a.** 100%

$150,000 increased to $300,000 is a 100% increase.

11. **d.** from 25% to 31%

$28\% - 3\% = 25\%$

$28\% + 3\% = 31\%$

12. **a.** $\frac{1}{4}$

$22\% + 3\% = 25\% = \frac{1}{4}$

13. **e.** $1,200

8% of $15,000 = 0.08 × $15,000 = $1,200

14. **c.** between 38% and 44%

15. **d.** $\frac{1}{5}$

21% is approximately 20%.

$\frac{20}{100} = \frac{1}{5}$

16. **d.** Administration and finance expenses accounted for almost $1 out of every 10 in the 2010 Massachusetts budget.

$9\% = \frac{9}{100}$ or approximately $\frac{10}{100}$ or $\frac{1}{10}$

17. **d.** $\frac{1}{2}$

51% is approximately 50% or $\frac{1}{2}$

18. **b.** 2 (40 ÷ 61 = 0.66)

19. **d.** The percent of students who buy their lunch at a nearby elementary school.

20. **e.** Of the fourth graders, 70% buy their lunch.

$\frac{35}{50} = \frac{70}{100} = 70\%$

21. **c.** 3

9.5 million − 6.5 million = 3 million

22. **b.** 5.1 to 13.5

23. **a.** The graph shows the projected number of Americans age 65 and older with Alzheimer's disease between 2010 and 2050.

24. **c.** The number of Americans age 65 and older with Alzheimer's disease is projected to more than double between 2010 and 2040.

25. **b.** See Perfect Contact Lenses Manufacturers

26. **b.** Would you prefer to wear soft and comfortable contact lenses or your glasses?

27. **c.** 1,000 people walking out of a contact lenses fitting office

28. **a.** 1 out of 10

29. **b.** noon − 3 P.M.

30. **e.** to show the percent of Best Beverage soft drinks sales during different time periods of the day

31. **e.** 12% of soft drink sales occur after 7 P.M.

32. **c.** just over 15

33. **a.** The graph shows the average hourly wages of selected jobs in May 2009.

34. **b.** Of the jobs listed, cashiers and fast food workers are the lowest hourly wage earners.

35. **d.** registered nurse

MEASUREMENTS

Time

365 days = 1 year

12 months = 1 year

52 weeks = 1 year

7 days = 1 week

24 hours = 1 day

60 minutes = 1 hour

60 seconds = 1 minute

Length and Area

5,280 feet = 1 mile

1,760 yards = 1 mile

3 feet = 1 yard

36 inches = 1 yard

12 inches = 1 foot

144 square inches = 1 square foot

4,840 square yards = 1 acre

1,000 meters = 1 kilometer

100 centimeters = 1 meter

1,000 millimeters = 1 meter

10 millimeters = 1 centimeter

Weight

2,000 pounds = 1 ton

16 ounces = 1 pound

1,000 grams = 1 kilogram

1,000 milligrams = 1 gram

Volume

4 quarts = 1 gallon

2 pints = 1 quart

4 cups = 1 quart

2 cups = 1 pint

32 ounces = 1 quart

16 ounces = 1 pint

8 ounces = 1 cup

1,000 milliliters = 1 liter

GLOSSARY

A

analyze To study or examine closely

average The middle value of a group of numbers. An average is found by adding a group of numbers and dividing the total by the number of items in the group.

Example: What is the average of these numbers: 35, 25, 48?

$35 + 25 + 48 = 108$ $108 \div 3 = \mathbf{36}$

axis One of the straight lines, either vertical or horizontal, that run from bottom to top or left to right on a graph

B

bar graph A visual picture of data that uses rectangles to represent number values

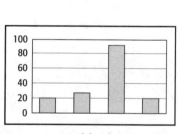
vertical data bars

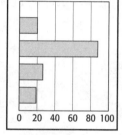
horizontal data bars

bias An attitude that always favors one way of thinking over another

C

chart A visual representation of data that uses columns and rows to make numbers easier to interpret

circle graph A visual representation of data that illustrates size or percent as pie-shaped parts of a circle

circle graph

column A vertical arrangement of numbers

compare To decide whether one number is equal to, less than, or greater than another number

Example: $3.7 > 3\frac{1}{2}$ (3.7 is greater than $3\frac{1}{2}$.)

conclusion A decision reached by critical thinking

convert To change from one unit of measurement to another

Example: How many feet are there in 60 inches?

$60 \div 12$ (inches per foot) $= \mathbf{5\ feet}$

correlation A mutual relationship between two things

D

data A collection of numbers that gives information about something

decimal A value less than 1 and written to the right of a decimal point

Example: 12.125 (0.125 is a decimal and is less than one.)

E

estimate To find an approximate number, a number that is close to the actual value

An estimate for $597 + 101$ would be $600 + 100 = 700$.

equation A mathematical statement that says two values are equal

Example: $3 + 8 = 11$ is an equation. $27 - x = 20$ is an equation.

expression The use of math symbols to represent the relationship between numbers

Example: $x \div 25$ (x is being divided by 25)

F

fraction A number less than 1, written as part over whole

Example: $\frac{3}{4}$ is a fraction (3 is the part; 4 is the whole).

friendly numbers Numbers that are easy to work with

Example: 100 and 10 are friendly numbers because one divides easily into the other.

H

heading A label that appears at the top of a graph or chart

horizontal The direction from left to right

K

key The part of a graph that provides an explanation or identification of symbols

L

label The words used to identify what the numbers on a graph or chart refer to

line graph A picture of data that uses lines or segments to connect numerical values

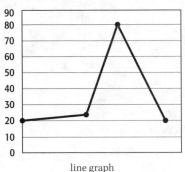

line graph

M

mean Average value; the amount found by adding a group of numbers and dividing by the number of items in the group

median The number in a series of numbers arranged from smallest to largest at which there is an equal number of values higher and lower

mode The most frequently appearing value in a set of data

multistep problem A problem that takes more than one operation to solve

O

order of operations The acceptable order in which to perform math operations in a multistep problem

Example: In the problem $(5 + 4) \div 3$, you add the numbers in parentheses before you divide.

ordering The process of putting numbers in order from least to greatest or greatest to least

P

percent A whole that is divided into one hundred equal parts. % = percent

Example: 43% means 43 out of 100.

pictograph A visual picture of data that represents values with small symbols

horizontal display

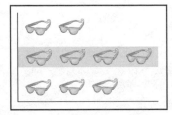

vertical display

place value The position a digit holds in a number

Example: In 27,439, the 7 is in the thousands place.

prediction A foretelling based on observation and analysis of data

probability A measure of how often an event is likely to happen

process A step-by-step method for completing a task

R

range The variation between the lowest and highest values in a set of data

rate A comparison of one number to another that does not change; a constant ratio

ratio A comparison of one number to another

Example: 4/5, 4:5, and 4 to 5 are all ratios comparing 4 to 5.

rounding Changing the value of a number slightly so that it ends in a number that is easy to work with

Example: 89 can be rounded to 90; 2.03 can be rounded to 2.

row A horizontal arrangement of numbers

S

scale The series of spaces on the axis of a graph that increase equally

set-up problem A problem that does not require computation. Instead, it requires you to choose the correct method for solving a problem.

source The place from which data is obtained

statistics The collection, organization, and interpretation of data

survey A scientific gathering of data

T

table A chart of data that shows rows and columns of data

trend A general direction that shows a consistent pattern

V

vertical The direction from top to bottom

W

word problem A math problem that asks a question in words

INDEX